APPLIED STATISTICAL MODELING AND DATA ANALYTICS

APPLIED STATISTICAL MODELING AND DATA ANALYTICS

A Practical Guide for the Petroleum Geosciences

SRIKANTA MISHRA
Battelle Memorial Institute, Columbus, Ohio, USA

AKHIL DATTA-GUPTA
Texas A&M University, College Station, Texas, USA

ELSEVIER

Elsevier
Radarweg 29, PO Box 211, 1000 AE Amsterdam, Netherlands
The Boulevard, Langford Lane, Kidlington, Oxford OX5 1GB, United Kingdom
50 Hampshire Street, 5th Floor, Cambridge, MA 02139, United States

Library of Congress Cataloging-in-Publication Data
A catalog record for this book is available from the Library of Congress

British Library Cataloguing-in-Publication Data
A catalogue record for this book is available from the British Library

ISBN: 978-0-12-803279-4

For information on all Elsevier publications
visit our website at https://www.elsevier.com/books-and-journals

Working together
to grow libraries in
developing countries

www.elsevier.com • www.bookaid.org

Publisher: Candice Janco
Acquisition Editor: Amy Shapiro
Editorial Project Manager: Tasha Frank
Production Project Manager: Anitha Sivaraj
Cover Designer: Christian J. Bilbow

Typeset by SPi Global, India

Dedication

This book is dedicated to

*our **parents**, for teaching us the letters and giving us the spirit;*
*our **spouses**, for sustaining the present with love and understanding;*
*and our **children**, for continuing the quest of knowledge.*

न हि ज्ञानेन सदृशं पवित्रमिह विद्यते ।

na hi jnaanena sadrsham pavitram iha vidyate

There is nothing as pure as knowledge.
Bhagavad Gita, IV.38

Contents

8. Data-Driven Modeling

9. Concluding Remarks

Companion site for this book is available at https://www.elsevier.com/books-and-journals/book-companion/9780128032794

Preface

As practicing petroleum engineers and geoscientists, we have long been fascinated by the myriad applications of statistical methods in characterizing, monitoring, and forecasting the behavior of subsurface fluid-bearing geosystems. Statistics has been a quintessential tool in our arsenal for fundamental tasks such as describing and modeling data from well logs, core samples, and injection/production tests. It has improved our designs of laboratory, field, and numerical experiments to understand the relationship between geologic inputs (e.g., porosity, permeability, and well-log attributes) and state variables (e.g., pressure and production rates). Statistical techniques have also helped improve estimates of uncertainty in static and dynamic reservoir model predictions arising from the underlying model input uncertainty. Our journey has taken us from basic exploratory data analysis and regression modeling to more advanced multivariate analysis, nonlinear and nonparametric regression modeling, experimental design and response surface analysis, and uncertainty quantification methods. Lately, we have been intrigued by the promise of big data analytics for oil and gas projects based on its success in other problem domains and have been exploring machine-learning techniques to develop data-driven insights for understanding and optimizing the performance of petroleum reservoirs.

During this "random walk" along the road of applied statistics, we have studied many research papers on the subject from fellow petroleum engineers and geoscientists and have also contributed to the growing literature. We have consulted several books on statistical modeling and data analytics that target both specialized and broad audiences. Through serendipity, we have come to realize that there is no single textbook or reference volume currently in the market that addresses the theory and practice of these topics from the perspective of petroleum engineering or geoscience applications.

This book is our humble attempt to fill the void. It seeks to provide a practical guide, via theoretical background and practical examples, to many of the classical and modern statistical techniques that have become, or are becoming, mainstream for oil and gas professionals. It is intended to serve as a "how to" reference for the practicing petroleum engineer or geoscientist interested in applying statistical modeling and data analytics techniques in formation evaluation, reservoir characterization, reservoir modeling and management, and production operations.

This is a book on the application of statistics, written by practitioners, for practitioners. As such, we have tried to strike a judicious balance between statistical rigor and formalism and practical considerations regarding the fundamentals and applicability of various relevant concepts. Beginning with a foundational discussion of exploratory data analysis (Chapter 2), probability distributions (Chapter 3), and linear regression modeling (Chapter 4), the book focuses on fundamentals and practical examples of such key topics as multivariate analysis (Chapter 5), uncertainty quantification (Chapter 6), experimental design and response surface analysis (Chapter 7), and data-driven modeling (Chapter 8). Datasets from the petroleum geosciences are extensively used to demonstrate the applicability of these techniques. We have chosen not to discuss topics related to geostatistics or time series analysis, as there are several excellent practical references available on the subject.

Although this book is primarily organized in the form of a ready reference guide for practitioners in the petroleum geosciences, it can also be used as a textbook for an upper division or graduate-level course on the subject. To that end, we have added several pedagogical examples and exercise problems to each chapter. The book will also be useful for professionals dealing with subsurface flow problems in hydrogeology, geologic carbon sequestration, and nuclear waste disposal. The material in the book has been collated from class notes of graduate and undergraduate courses that we have taught and short courses and workshops that we have offered at professional society meetings and client locations.

We view statistical modeling and data analytics as both an established and an emerging field, where basic concepts from classical statistics provide the building blocks for applying newly developed algorithms from the computer and data science domains. We hope this book will empower petroleum engineers and geoscientists with a greater appreciation of relevant principles and tools for converting data into information—particularly the actionable kind that lead to better decisions.

Acknowledgments

We wish to express our gratitude and appreciation to

- our professors at Stanford University, The University of Texas at Austin, and Indian School of Mines—who taught us how to formulate and solve problems of practical significance; shared with us their knowledge, experience, and insights; and inspired us with their scholarship and professionalism;

- our students in university classes and professional short courses on statistical modeling and uncertainty analysis—who beta tested the class notes and slides that morphed into this book and frequently challenged us to refine our understanding of the subject matter;

- our employers Battelle Memorial Institute and Texas A&M University and multiple governmental and industrial sponsors—who provided us with the resources for carrying out much of the research that culminated in this book;

- our colleague Jared Schuetter—who contributed significantly to Chapters 7 and 8; and helped improve our understanding of big data analytics;

- our colleagues Banda RamaRao, Neil Deeds, David Sevougian, and Mike King—who collaborated in many enriching topical discussions and contributed to this book through concepts, text, and figures;

- our colleagues Shuvajit Bhattacharya and Isis Fukai and advisees Atsushi Iino, Changdong Yang and Aditya Vyas—who reviewed the draft manuscript from the perspective of practicing petroleum engineers and geoscientists and provided many helpful comments;

- our Elsevier editors Tasha Frank and Louisa Hutchins and Production Manager Anitha Sivaraj—who guided us through the proposal development, manuscript submission, and book production processes with patience; and last, but not the least,

- the Mishra family (Snigdha, Ananya, and Anshuman) and the Datta-Gupta family (Mausumi, Alina, and Antara) who indulged this passion of ours by willingly giving us the gift of time.

CHAPTER

1

Basic Concepts

1.1 BACKGROUND AND SCOPE

We introduce the reader to some fundamental concepts of classical statistics such as probability and random variables, along with basic concepts from the emerging field of data analytics and big-data technologies. We also list some typical applications of the relevant techniques for data analysis in the petroleum geosciences.

1.1.1 What Is Statistics?

Statistics is the science of acquiring and utilizing data. It provides us with the tools for data collection, summarization, and interpretation—with the goal of identifying the underlying structure, trends, and relationships inherent in the data. This is how we convert data into information.

Fundamental to statistics are the concepts of *population* and *sample*. A population is the universe of all possible outcomes and events, whereas a sample is a finite subset extracted

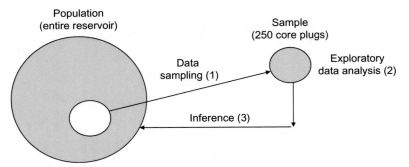

FIG. 1.1   **Schematic showing population—sample relationship.**

from the population. Statistical analyses are performed on the sampled data to draw *inference* about the characteristics of the population, without having to study the entire population. The population is exhaustive and is characterized by its *parameters*. The sample is limited and is characterized by the *statistic* that is related to the population parameters.

Fig. 1.1 shows a schematic of the relationship between population and sample. Here, the population represents permeability values for an entire oil reservoir at the scale of a small core plug. To learn more about the distribution of permeability values, in step (1), we randomly sample this population using a finite number of core plugs (e.g., 250). In step (2), we analyze these permeability values to determine the proportion of plugs with permeability greater than 10 mD (e.g., 65%). Finally, in step (3), we determine the representativeness of this result for the entire population (e.g., 95% certain that margin of error is ±6%).

Application of statistics to any dataset generally begins with exploratory data analysis. Here, the goal is to quantify and visualize the range of values a given variable can take, summary attributes such as averages and spread, and the nature and strength of correlation between two or more variables (Chapter 2). In the next step, the distribution of the variable is examined to understand the relative likelihood of various values within the observed range and the possibility of describing the distribution using a compact mathematical form (Chapter 3). Another common task involves exploring how the relationship between two variables can be described using a linear regression model or variants thereof (Chapter 4). When multiple variables are included in the dataset, it is useful to identify the degree of redundancy among different variables and if the dataset can be partitioned into any statistically homogeneous subpopulations (i.e., clusters). This is the scope of multivariate analysis (Chapter 5).

The broad classes of techniques described above fall within the realm of classical statistics and have been employed by petroleum engineers and geoscientists for many years (see Stanley, 1973 and references therein). Recent contributions (e.g., Davis, 2002; Jensen et al., 2000) discuss the geoscience-oriented application of these techniques in greater detail, including other topics not covered in this book such as geostatistics and time series analysis.

Statistical methods are also relevant in the context of uncertainty analysis, where the goal is to translate the uncertainty in the inputs of a model into uncertainty in corresponding model predictions (Chapter 6). Here, the concepts mentioned in the previous paragraph are fundamental to characterizing the uncertainty both in the model inputs and the model results and building predictive models that relate the specified uncertain inputs to the computed uncertain outputs. Another important application is with respect to design of experiments, both physical and computational (Chapter 7). Statistical approaches are useful for determining how to construct a limited number of experiments that properly span the design space and how to fit a response surface to the experimental results that can be used as a surrogate model.

1.1.2 What Is Big Data Analytics?

The terms "big data" and "data analytics" have become quite the buzzword in recent years, especially because of many reported applications in areas such as consumer marketing, health and life sciences, and national security. This has led to the perception that big data analytics has the potential to be a game changer for oil and gas applications (Holdaway, 2014). The industry is beginning to explore the possibilities of "mining" large volumes of data about the subsurface, physical infrastructure, and flows to obtain new insights about the reservoir that can help increase operational efficiencies.

Big data generally refers to large, multivariate datasets characterized by the three V's: volume, variety, and velocity (Fig. 1.2). *Volume* refers to the size of the data, where we are increasingly dealing with $\sim 10^2$–10^4 independent variables and $\sim 10^3$–10^6 observations or data records, each collected at multiple temporal and/or spatial locations. *Variety* refers to data in multiple formats such as numbers, video, and text, which can be both structured and unstructured, and requires a combination of numerical methods, image analysis, and/or natural language processing. *Velocity* refers to the growing ubiquity of real-time streaming data from downhole sensors or surface gauges, which adds to the size of the dataset with additional considerations such as data archival, resampling, and redundancy analysis.

As shown in Fig. 1.2, *data analytics* is the process of (a) examining the data, (b) understanding what the data say and "learning" from the data, and (c) making predictions based on these data-driven insights that (hopefully) lead to better decisions (Hastie et al., 2008). Essentially, data analytics methods are applied to help understand hidden patterns

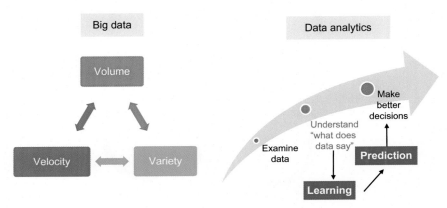

FIG. 1.2 **Big data analytics—what and why.**

and relationships in large and complex datasets. A number of equivalent terms such as *statistical learning*, *knowledge discovery*, *data mining*, and *data-driven modeling* are often interchangeably used to describe this collection of techniques, which are drawn from computer science, machine learning, and artificial intelligence (Chapter 8).

From an information technology perspective, however, the scope of data analytics is somewhat broader because it includes the following steps (IDC Energy Insights, 2014):

- *Data organization and management*, which involves data collection, warehousing, tagging, QA/QC, normalization, integration, and extraction.
- *Analytics and discovery*, which involves software-driven analysis, predictive model building, and extraction of data-driven insights.
- *Decision support and automation*, which involves deploying rule-based systems with functionality to support collaboration, scenario evaluation, and risk management.

Although big data has not become ubiquitous in the oil and gas industry, a vision for how big-data-related technologies can be implemented in the context of exploration and production operations is described in Brulé (2015).

1.1.3 Data Analysis Cycle

For petroleum geoscience applications, it is more useful to consider statistical modeling and data analytics as part of an integrated data analysis cycle as shown in Fig. 1.3. The scope of various work elements that comprise this cycle are explained below.

Data collection and management. This step involves the acquisition and aggregation of data from multiple sources (e.g., cores, well logs, and production records), possibly in multiple forms (e.g., numbers and text). The data also undergo a QA/QC process to ensure the traceability and accuracy of each data record. Finally, the data have to be made easily available for visualization and analysis.

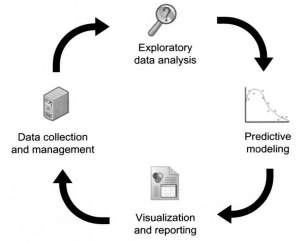

FIG. 1.3   **Schematic of data analysis cycle.**

Exploratory data analysis. The goal of this step is to develop a preliminary understanding of the data in terms of the characteristics of individual variables and the relationship among various variables. Other objectives include identifying key variables of interest, formulating questions for digging deeper into the data, and selecting techniques that will be used for detailed analysis. The relevant concepts involved in this step are discussed in Chapters 2 and 3.

Predictive modeling. The analyses in this step generally begin with *unsupervised learning*, where the issues of redundancy among the independent variables and possible reduction in data dimensionality (without losing any information) are first addressed. This is followed by *supervised learning*, where observed values of a response variable are used to train a model between the independent variables (i.e., predictors) and the dependent variable (i.e., response). This predictive model can then be used to answer questions posted in the previous step. Chapters 4–8 discuss the relevant concepts that are integral to this step.

Visualization and reporting. The ultimate goal of any modeling and/or analysis is to provide input for a decision by transferring information to decision-makers. It is therefore necessary to capture what has been learned in the form of visual summaries, compact reports, or decision-support tools that can be used to answer "what-if" type questions. Another useful outcome from this step is the use of insights from predictive modeling to identify what new data should be collected and the kinds of questions to pursue in the future.

1.1.4 Some Applications in the Petroleum Geosciences

The principles described throughout the book are explained with the help of many illustrative examples and problems to demonstrate their practical applicability. These include the following:

- Determining conditional probabilities of cause-effect relationships
- Computing summary statistics (e.g., mean and variance)
- Calculating correlation and rank correlation coefficients between two variables
- Visualizing univariate, bivariate, and multivariate data
- Estimating probability coverage levels for different distributions
- Analyzing behavior of normal and lognormal distributions
- Calculating confidence interval and sampling distribution for the mean
- Testing for significance of difference in means
- Comparing two different distributions for statistical equivalence
- Fitting simple and multiple linear regression models to observed data
- Developing a nonparametric regression model from given data
- Reducing data dimensionality with principal component analysis
- Grouping data with k-means and hierarchical clustering
- Identifying classification boundary between clusters using discriminant analysis
- Developing distributions from data, limited knowledge, or subjective judgment
- Translating model input uncertainty into uncertainty in model predictions using Monte Carlo simulation and analytic alternatives

- Analyzing input-output dependencies from Monte Carlo simulation results
- Creating an experimental design and fitting a response surface to the results
- Applying machine learning techniques (e.g., random forest, gradient boosting machine, support vector regression, and kriging model) for predictive modeling
- Generating decision rules with classification tree analysis

Some of the examples listed here are purely pedagogic in nature, while others are based on actual datasets (albeit reduced in size to make the presentation tractable). Finally, several field datasets have been analyzed to demonstrate how multiple methods "come together" in the context of linear and nonparametric regression analysis, multivariate analysis, and data-driven modeling.

1.2 DATA, STATISTICS, AND PROBABILITY

1.2.1 Outcomes and Events

Generally, there is some degree of unpredictability or randomness associated with most natural phenomena. We can represent this unpredictability in terms of the many possible outcomes of an experiment to define "what can happen." Simply put, statistics is concerned with the determination of the probable (events) given the possible (outcomes) (Davis, 2002). Formally stated, *outcomes* are elements of the sample space Ω, *events* are an appropriate subset of Ω, and *probability*, P, is the likelihood of the event occurring ($0 \leq P \leq 1$).

The sample space, Ω, is a set whose elements describe outcomes of the experiment of interest. For example, if the experiment is a wildcat well with two possible outcomes—dry well (D) or success (S), then the sample space is $\Omega = \{D, S\}$. If the experiment is porosity determination from core samples with multiple possible outcomes (equal to the number of samples), then the sample space is $\Omega = \{0, 1\}$. Another experiment could be the order in which three wells are tested—leading to six different outcomes—with the sample space being $\Omega = \{123, 132, 213, 231, 312, 321\}$.

Events are subsets of the sample space, that is, event A occurs if the outcome of the experiment is an element of set A. For example, let A be the event where well #1 is tested either first or second, that is, $A = \{123, 132, 213, 312\}$. Similarly, let B be the event where well #2 is tested either first or second, that is, $\{123, 213, 231, 321\}$. When both A and B occur, we refer to this as *intersection*, symbolically denoted as $A \cap B = \{123, 213\}$. If at least one of A or B occurs, we refer to this as *union*, symbolically denoted as $A \cup B = \{123, 132, 213, 312, 231, 321\}$. The *complement* of A, denoted by A^C, is when A does not occur, that is, $A^C = \{231, 321\}$. Note that the complement of the sample space Ω is the null set Φ. A and B are considered to be *disjoint* (i.e., mutually exclusive) if there are no common elements, that is, $A \cap B = \Phi$. Some additional results follow from *De Morgan's law*, which states that $(A \cup B)^C = A^C \cap B^C$ and $(A \cap B)^C = A^C \cup B^C$. The concepts of intersection, union, and complement are schematically shown below in Fig. 1.4.

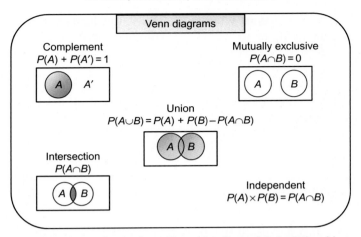

FIG. 1.4   **Concepts of intersection, union, and complement and associated relationships.**

1.2.2 Probability

Probability is the likelihood of an event occurring and is expressed as a number between 0 and 1 or equivalently as a percentage between 0% and 100%. In the *frequentist* view of probability, it is the relative frequency with which an event occurs in a long sequence of trials and is based on historical or measured data. In the *Bayesian* (subjective) view, it is the degree of subjective belief about the event occurring given all relevant information.

As an example of the frequentist approach, if observed net pay (in ft) from nine wells are $h=$ [17.5, 20.4, 15.6, 16.2, 16.9, 18.3, 9.4, 15.2, 18.3], then the probability that the net pay is greater than 18 ft is given by $P[h \geq 18] = 4/9 = 0.44$. On the other hand, the statement "based on prior evidence and expert judgment, the probability is 30% that at least one well will exceed 1000 BOPD in initial production" is an example of the subjective (Bayesian) approach.

Historically, the early use of probability was in the subjective vein. Pierre Bernoulli (1713), Thomas Bayes (1763), and Pierre Laplace (1812) treated probability as plausibility, given all evidence. It was only in the mid-19th century that mathematicians started considering probability as the long-run relative frequency and as an objective tool, based on data, for dealing with random phenomena. This led to the development of statistics as an independent branch of mathematics. In the mid-20th century, the information theoretician Edwin Jaynes (1957) promoted the application of the Bayesian framework as a formal basis for conditioning probabilities. In this book, we embrace both the frequentist view and the subjective view of probability and use the formalism that is most appropriate for the amount of data and the problem at hand.

Some basic rules governing probability are discussed next.

(A) Total probability of the sample space is unity

$$P(\Omega) = P(A) + P(A^C) = 1 \tag{1.1}$$

(B) The probability of the union of two events (as per Fig. 1.4) is given by

$$P(A \cup B) = P(A) + P(B) - P(A \cap B) \tag{1.2}$$

(C) For mutually exclusive events, where $P(A \cap B) = 0$, this leads to the *additivity rule* that states that the probability of an event is the sum of the probability of mutually exclusive outcomes belonging to that event:

$$P(A \cup B) = P(A) + P(B) \tag{1.3}$$

(D) For independent events, where the experiments do not influence each other, the probabilities are multiplicative, that is,

$$P(A \cap B) = P(A) * P(B) \tag{1.4}$$

As an example, consider a five-well wildcat campaign with outcomes 1 (success) and 0 (failure). We are interested in the event where exactly one well was a success. This can be enumerated as

$$A = \{(0,0,0,0,1), (0,0,0,1,0), (0,0,1,0,0), (0,1,0,0,0), (1,0,0,0,0)\}.$$

The probability for each element of A is clearly $p(1-p)^4$, where p is the probability of success. Hence, the probability of event A can be written as

$$P(A) = 5p(1-p)^4$$

We can generalize this to state that the probability of r successes in n trials, when the probability of success in a single trial is p, is given by

$$P = {}^nC_r p^r (1-p)^{n-r}$$

1.2.3 Conditional Probability and Bayes Rule

Let event A, with probability $P(A)$, lead to event B, with probability, $P(B)$. We denote $P(B|A)$ as the *conditional probability* of event B, given that event A has occurred. If A and B are independent events, that is, event B does not depend on event A, then $P(B|A) = P(B)$. The concept of conditional probability can be explained in terms of the intersection of two events (as shown in Fig. 1.4), by noting that $P(B|A)$ is simply the fraction of probability of A that is also in event B. In other words,

$$P(B|A) = P(A \cap B)/P(A) \tag{1.5}$$

This leads to the *multiplication rule* for the probability of the intersection of two events, which is also a statement of symmetry in expressing conditional probabilities:

$$P(A \cap B) = P(B|A) * P(A) = P(A|B) * P(B) \tag{1.6}$$

A related concept is *total probability*, which is based on the computation of probabilities by considering all disjoint events that belong to a sample set. Consider the example shown in Fig. 1.5, where C_1, C_2, and C_3 are disjoint events that collectively make up the sample set Ω (indicated by the rectangle) and A is another event that belongs to Ω (indicated by the filled circle).

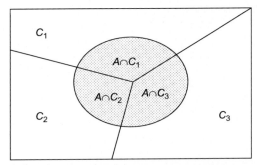

FIG. 1.5   Concepts of intersection, union, and complement in the context of conditional probability.

Another way of thinking about these relationships is to denote $C_j \equiv$ cause and $A \equiv$ effect. Using the additivity rule, the probability of A can be written as

$$P(A) = P(A \cap C_1) + P(A \cap C_2) + P(A \cap C_3) \tag{1.7}$$

From the definition of conditional probabilities in Eq. (1.6), we have

$$P(A \cap C_j) = P(A|C_j) * P(C_j) \tag{1.8}$$

Substituting in Eq. (1.7), we get

$$P(A) = P(A|C_1) * P(C_1) + P(A|C_2) * P(C_2) + P(A|C_3) * P(C_3) \tag{1.9}$$

A formal linkage between the probabilities of C_j and A can be compactly established using *Bayes' rule*, as discussed next. Using the identities in Eqs. (1.5) and (1.8), we can write

$$P(C_j|A) = \frac{P(A|C_j) * P(C_j)}{P(A)} = \frac{P(A|C_j) * P(C_j)}{\sum_j P(A|C_j) * P(C_j)} \tag{1.10}$$

This is the basic statement of Bayes' rule, which can be restated as follows:

$$P(\text{cause}_j|\text{effect}) = P(\text{effect}|\text{cause}_j) * P(\text{cause}_j)/P(\text{effect}) \tag{1.11}$$

where $P(\text{effect})$ is simply a normalizing constant. Thus, Bayes' rule allows us to make inferences about possible causes, given observed effects, starting with the information regarding the probability of different effects from each possible cause. It enables us to combine the information content of the data with our prior knowledge to obtain a more refined statistical distribution. This turns out to be a very powerful tool for updating knowledge in an objective manner.

Consider the following illustrative example. We are interested in understanding the causes of poor well productivity (i.e., wells with initial production less than 100 barrels/day) in a fractured reservoir, denoted as event A. Let B_1 denote the event where the well-test permeability is greater than 100 mD and B_2 denote the event where the well-test permeability is less than 20 mD. From operational records, we know that $P(B_1) = 0.6$, and $P(B_2) = 0.40$. Furthermore, from production data, we know that poor well productivity is more likely to occur in low-permeability conditions, such that $P(A|B_1) = 0.07$ and $P(A|B_2) = 0.95$. If we drill a new well and encounter poor well productivity, what is the probability that we are in a low-permeability environment?

TABLE 1.1   An Example Tabular Calculation Using Bayes' Rule

Effect A	Cause B_i	$P(B_i)$	$P(A|B_i)$	Product	Scaled Probability $P(B_i|A)$
Low well productivity	$k > 100$ mD	0.6	0.07	0.042	0.10
	$k < 20$ mD	0.4	0.95	0.38	0.90
Sum				0.422	

We start by calculating the total probability using Eq. (1.9) as follows:

$$P(A) = P(A|B_1) * P(B_1) + P(A|B_2) * P(B_2) = 0.07 * 0.6 + 0.95 * 0.4 = 0.422$$

Next, we calculate the probability of a low-permeability reservoir (event B_2), given poor well productivity (event A), using Eq. (1.10) as follows:

$$P(B_2|A) = P(A|B_2) * P(B_2)/P(A) = 0.95 * 0.4/0.422 = 0.9$$

The *prior* probability for low-permeability conditions was $P(B_2) = 0.40$. The *posterior* probability for low-permeability conditions, given low well productivity, is now $P(B_2|A) = 0.90$. In other words, the knowledge of poor well productivity has significantly improved our confidence in identifying low-permeability conditions. A tabular format for performing these calculations is presented below in Table 1.1.

1.3 RANDOM VARIABLES

1.3.1 Discrete Case

A random variable (RV) is a quantity whose value is subject to variations due to randomness. Therefore, RVs can have many possible values, which can be either discrete or continuous. For example, the number of downhole gauge failures in a given month is a discrete RV, whereas porosity values obtained from core analysis in a given well can be treated as a continuous RV.

The probability mass function (PMF), p, of a discrete RV, X, denotes the probability that the RV is equal to a specified value, a. This is denoted by

$$p(a) = P(X = a) \tag{1.12}$$

Similarly, the cumulative distribution function (CDF), F, denotes the probability that X will take on values equal to or less than a. Symbolically, this is represented as

$$F(a) = P(X \leq a) = \sum_i p(a_i) \text{ with } a_i \leq a \tag{1.13}$$

Consider the case of two die throws, where we are interested in tracking the maximum value from each die. We can enumerate the possible outcomes in a tabular form, as shown in Table 1.2.

Fig. 1.6 shows the PMF and CDF for this example.

TABLE 1.2   Example of a Discrete RV

a	1	2	3	4	5	6
$p(a)$	1/36	3/36	5/36	7/36	9/36	11/36
$F(a)$	1/36	4/36	9/36	16/36	25/36	1

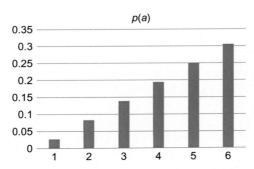
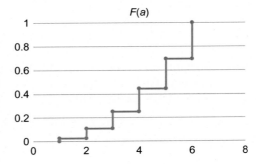

FIG. 1.6   **Example PMF and CDF for a discrete RV.**

1.3.2 Continuous Case

An RV is defined to be continuous, if for some function, f, and any two numbers a and b with $a \leq b$:

$$P(a \leq X \leq b) = \int_a^b f(x)dx \tag{1.14a}$$

$$f(x) \geq 0 \ \text{ for all } x \ \text{ and } \int_{-\infty}^{+\infty} f(x)dx = 1 \tag{1.14b}$$

Here, f is the probability density function (PDF), which is the continuous version of the PMF. Similarly, we can define the cumulative distribution function (CDF), F, for the continuous case as follows:

$$P(a < X \leq b) = P(X \leq b) - P(X \leq a) = F(b) - F(a) \tag{1.15a}$$

$$F(b) = \int_{-\infty}^b f(x)dx \tag{1.15b}$$

$$f(x) = \frac{d}{dx}F(x) \tag{1.15c}$$

We discuss the nature of discrete and continuous RVs in greater detail in Chapter 3.

1.3.3 Indicator Transform

An indicator transform provides a mechanism to transform a continuous random variable to a discrete random variable. For example, properties such as porosity and permeability in a reservoir are continuous variables. Such continuous variables can be represented in terms of discrete indicator variables by introducing thresholds or cutoffs. A common application is a geologic facies definition that might be based on cutoffs introduced for specified properties.

The indicator transform associated with random variable X for a threshold value x_k is defined as follows:

$$I(x_k; X) = \begin{cases} 0 & \text{if } X > x_k \\ 1 & \text{if } X \leq x_k \end{cases} \tag{1.16}$$

An important property of an indicator random variable is that its expected value, $E\{I(x_k; X)\}$, is equal to the cumulative probability, $P(X \leq x_k)$, that is, the proportion of X below x_k. This can be easily seen below:

$$E\{I(x_k; X)\} = 1 \times P(X \leq x_k) + 0 \times P(X > x_k) = P(X \leq x_k) \tag{1.17}$$

1.4 SUMMARY

In this chapter, we began with some introductory text regarding statistics and the statistical modeling process, followed by a discussion of similar concepts related to big-data analytics and data analysis cycle. We also presented an overview of fundamental statistics and probability terms, conditional probability, and random variables. Our goal here was to provide the foundational concepts that will be used as building blocks for the chapters to follow.

Exercises

1. What are some of the attributes that would make samples representative of the population in petroleum geoscience applications?
2. Using the OnePetro database, find one example of big-data application in each of the following areas: (a) drilling, (b) formation evaluation, (c) production, (d) reservoir management, and (e) predictive maintenance.
3. A biased coin is being tossed. The probability of getting heads is 0.51 and of getting tails is 0.49. We are interested in the number of tosses it takes until a head occurs for the second time. What is the probability that it takes five tosses?
4. If events E_1 and E_2 are independent with probabilities $P(E_1) = 0.4$ and $P(E_2) = 0.7$, find the probability of the following: (i) $P(E_1 \cap E_2)$; (ii) $P(E_1 \cap E_2^c)$; (iii) $P(E_1^c \cap E_2)$?
5. Geoscientists have postulated two structural models for the basin where your company is drilling exploratory wells. The probability of finding oil is 0.7 for the first model, and 0.2 for the second model. The likelihood of the first model being true is 0.4, whereas that of the second model is 0.6. If the first exploratory well strikes oil, what are the revised likelihoods of the two models?

6. Experience has shown that students are unable to submit their homework on time (NH) for one of two reasons: computer crash (CC) or dog eating the homework (DH). The probability of CC is known to be 0.20, with the probability of no-homework submission because of CC being 0.50. The probability of DH is known to be 0.01, with the probability of no-homework submission because of DH being 0.99. If a student was unable to submit the homework on time, what is the probability that a dog ate the homework?

7. If two dies are being thrown and their sum is the discrete random variable of interest, calculate and plot the PMF and the CDF.

8. Consider the PDF given by $f(x) = 0$ for $x \leq 0$ or $x \geq 1$, and $f(x) = 1/\sqrt{x}$ for $0 < x < 1$. What is the probability that X belongs to the interval $[10^{-3}, 10^{-1}]$?

References

Brulé, M.R., 2015. The Data Reservoir: How Big Data Technologies Advance Data Management and Analytics in E&P. Society of Petroleum Engineers, Richardson, TX. https://doi.org/10.2118/173445-MS.

Davis, J.C., 2002. Statistics and Data Analysis in Geology, third ed. John Wiley & Sons, New York, NY.

Hastie, T., Tibshirani, R., Friedman, J.H., 2008. The Elements of Statistical Learning: Data Mining, Inference, and Prediction. Springer, New York.

Holdaway, K.R., 2014. Harnessing Oil and Gas Big Data With Analytics. John Wiley & Sons, Hoboken, NJ.

IDC Energy Insights, 2014. https://www.hds.com/en-us/pdf/training/hitachi-webtech-educational-series-big-data-in-oil-and-gas.pdf.

Jensen, J., Lake, L.W., Corbett, P.W.M., Goggin, D., 2000. Statistics for Petroleum Engineers and Geoscientists. Elsevier, New York, NY.

Stanley, L.T., 1973. Practical Statistics for Petroleum Engineers. Petroleum Publishing Company, Tulsa, OK.

Exploratory Data Analysis

Exploratory data analysis, which is concerned with summarizing and visualizing data as a starting point for more detailed analyses, is the subject of this chapter. We restrict ourselves to numerical data (as opposed to text or images) and note that: (a) data can be univariate or multivariate, (b) data can be categorical or numerical, (c) random variables can have more than one value, and (d) distributions capture the values taken by variables, and the frequency with each specific value occurs.

2.1 UNIVARIATE DATA

Whether we are dealing with a population or a sample, the observed values of a variable are likely to be different from each other. It is useful to explore this intrinsic variability for a single variable numerically using measures that quantify the "average" value, the spread around this average, and the overall degree of asymmetry over the full range of observed values. These univariate measures are described below, along with some common graphical methods for visually examining and summarizing the data.

2.1.1 Measures of Center

For a random variable X, where x_i are the individual outcomes, the most common measure of central tendency is the *mean* or the expected value, defined as

$$E[X] = \bar{X} = \sum_{i=1}^{N} f_i x_i = \frac{1}{N} \sum_{i=1}^{N} x_i \tag{2.1}$$

where f_i is the relative frequency and is often assumed to be the same (i.e., $1/N$) for each sample. The mean (i.e., the arithmetic mean) is the weighted average of all values, based on the relative frequency. Two other useful measures of central tendency are (a) *mode*, which is the most likely (frequently occurring) value, and (b) *median*, which is the midpoint of the distribution.

Consider the 10-sample net pay data: h (ft), [13, 17, 15, 23, 27, 29, 28, 27, 20, 24]. Here, the mean is 21.3, mode is 27, and median is 21.5 (i.e., average of 20 and 23). The mean, median, and mode generally coincide for symmetrical (or near-symmetrical) distributions but can be very different if the distribution is asymmetrical. The mean is strongly impacted by the extreme values, whereas the median is more robust and less sensitive to the outliers.

Fig. 2.1 illustrates this schematically for two cases, where the median lies between the mode and the mean, but the mean and mode switch places depending on the nature of asymmetry (i.e., left-skewed or right-skewed). Also shown therein is the case of a distribution with two distinct modes. Generally, such *bimodal* (or, by extension, multimodal) behavior is indicative of the fact that the dataset is not statistically homogeneous—most likely because of the "mixture" of two or more distinct distributions. An example would be the combination of porosity data from two different lithofacies with significantly different characteristics.

Other commonly used averages beyond the arithmetic mean are the harmonic and geometric means. The *harmonic mean* is the reciprocal of the arithmetic mean of the reciprocals, that is,

$$\bar{X}_H = N \bigg/ \sum_{i=1}^{N} \frac{1}{x_i} \tag{2.2}$$

The *geometric mean* is the Nth root of the product of N observations, that is,

$$\bar{X}_G = (x_i x_2 \ldots x_N)^N \tag{2.3a}$$

$$\bar{X}_G = \exp\left[\ln\left(\bar{X}_G\right) \right] = \exp\left[\frac{1}{N} \sum_{i=1}^{N} \ln(x_i) \right] \tag{2.3b}$$

Table 2.1 presents a set of 21 data points of core-derived porosity values from the oil producing Rose Run sandstone in Ohio, the United States (POR_TAB2-1.DAT)* along with the calculation of the arithmetic, harmonic, and geometric means.

The harmonic mean can also be identified with the concept of resistances in series, whereas the arithmetic mean can be identified with resistances in parallel. Thus, the effective permeability of a stratified reservoir where the layers are parallel to each other would be given by the arithmetic mean. On the other hand, the effective permeability of a core holder, containing multiple core samples in series, would be the harmonic mean. For a system where

---

*This refers to the name of a data file that is available in the online resource section of the book.

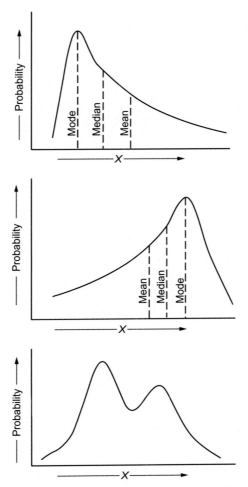

FIG. 2.1   **Location of mode for different types of distributions: (top) left-skewed, (middle) right-skewed, and (bottom) bimodal.**

permeability values vary in a random manner (as might be expected in the field), the effective value will be somewhere between these two bounds. As noted by Jensen et al. (2000), the effective permeability obtained from a pressure buildup test in the field is well approximated by the geometric mean, which falls between the arithmetic and harmonic means.

In geologic media averaging of properties, in particular permeability, can have important implications on the flow response. This is illustrated in Fig. 2.2 for basic and directional averaging of permeability in a sand-shale medium (AVG_FIG2-2.DAT). In general, the following basic averaging results hold:

Harmonic $\leq$ Geometric $\leq$ Arithmetic

and for 2-D/3-D directional averaging,

Harmonic $\leq$ Harmonic $-$ Arithmetic $\leq$ Arithmetic $-$ Harmonic $\leq$ Arithmetic

18

TABLE 2.1   Example Calculation of Various Measures of Central Tendency

	ϕ	$1/\phi$	$\ln(\phi)$
	8.1	0.123	2.092
	11.0	0.091	2.398
	13.0	0.077	2.565
	7.4	0.135	2.001
	6.5	0.154	1.872
	8.9	0.112	2.186
	6.5	0.154	1.872
	4.1	0.244	1.411
	7.9	0.127	2.067
	6.7	0.149	1.902
	11.0	0.091	2.398
	10.0	0.100	2.303
	9.1	0.110	2.208
	5.2	0.192	1.649
	3.1	0.323	1.131
	13.0	0.077	2.565
	12.0	0.083	2.485
	9.9	0.101	2.293
	9.5	0.105	2.251
	9.6	0.104	2.262
	9.3	0.108	2.230
Sum	181.8	2.760	44.140
Arithmetic mean	8.657		
Harmonic mean		7.608	
Geometric mean			8.182

Thus, some averages alter sand quality, whereas others preserve barriers to flow or flow around the barriers. For example, arithmetic averaging of sand and shale will behave more like sand, whereas harmonic averaging will behave more like shale. As shown by King et al. (1998), upscaling of reservoir properties specifically addresses this issue of averaging fine-scale geologic models to coarse-scale models.

2.1.2 Measures of Spread

The most fundamental measure of spread is the *variance*, which measures dispersion or variability around the mean. It is defined as

$$V[X] = \sigma_x^2 = \sum_{i=1}^{N} f_i(x_i - E[X])^2 = \frac{1}{N}\sum_{i=1}^{N}(x_i - E[X])^2$$
$$= \frac{\sum x_i^2}{N} - (E[X])^2 = E[X^2] - (E[X])^2$$

(2.4)

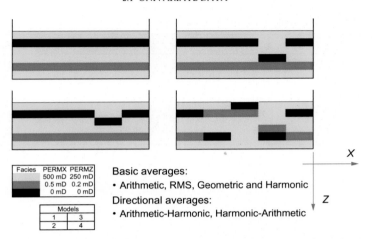

Basic averages:
• Arithmetic, RMS, Geometric and Harmonic

Directional averages:
• Arithmetic-Harmonic, Harmonic-Arithmetic

Facies	PERMX	PERMZ
	500 mD	250 mD
	0.5 mD	0.2 mD
	0 mD	0 mD

Models	
1	3
2	4

PERMX Averages	Model 1	Model 2	Model 3	Model 4
Arithmetic	333.42	333.42	333.42	333.42
Geometric	0.00	0.00	0.00	0.00
Harmonic	0.00	0.00	0.00	0.00
X-Direction Arithmetic-Harmonic	333.42	333.42	333.42	333.42
X-Direction Harmonic-Arithmetic	333.42	250.08	250.08	167.08

PERMZ Averages	Model 1	Model 2	Model 3	Model 4
Arithmetic	166.70	166.70	166.70	166.70
Geometric	0.00	0.00	0.00	0.00
Harmonic	0.00	0.00	0.00	0.00
Z-Direction Arithmetic-Harmonic	0.00	1.19	1.19	103.56
Z-Direction Harmonic-Arithmetic	0.00	0.00	0.00	0.00

FIG. 2.2   Basic and directional averaging of permeability in sand-shale media (King et al., 1998).

In other words, the variance is simply the difference between the mean of the squares and the square of the mean. The *standard deviation*, σ, is the square root of the variance and is also equivalent to the commonly used root-mean-square error (RMSE). The *coefficient of variance, CV*, is a normalized measure of spread, generally expressed as a percentage, and defined as

$$CV[X] = \frac{\sigma_x}{E[X]} \times 100\% \tag{2.5}$$

When quantifying heterogeneity in reservoir properties (e.g., permeability), the CV is a more consistent measure compared with variance or standard deviation as it focuses on the spread of the variable irrespective of the magnitude of actual values. A related measure in petroleum engineering is the Dykstra-Parsons coefficient assuming lognormal distribution of permeability, as discussed later in Chapter 3.

Note that the definitional equation for variance, Eq. (2.4), is actually the variance of a population. The variance of a finite sample, denoted by s, is obtained from Eq. (2.4) by replacing

TABLE 2.2   **Example Calculation of Variance**

	x	x^2	$(x - E[X])^2$
	8.1	65.61	1.21
	11.0	121.00	3.24
	13.0	169.00	14.44
	7.4	54.76	3.24
	6.5	42.25	7.29
Sum	46.0	452.62	29.42
$E[X]$	9.2		
$V[X]$	$=(452.62 - (46)^2/5)/4 = 29.42/4 = 7.355$		
$SD[X]$	2.712		

N in the denominator with $(N-1)$. This reflects the degrees of freedom available for calculating the variance, since one calculation is needed to calculate the mean. The modified equation is

$$s_x^2 = \frac{1}{N-1} \sum_{i=1}^{N} (x_i - E[X])^2 = \frac{1}{N-1} \left\{ \sum x_i^2 - \frac{\left(\sum x_i \right)^2}{N} \right\} \tag{2.6}$$

Table 2.2 shows the calculation of variance using the first five values of porosity from Table 2.1.

2.1.3 Measures of Asymmetry

In general, we can describe the variability of a dataset using moments around the mean or central moments, defined as

$$\mu_n = E[(X - \mu)^n] = \int_{-\infty}^{\infty} (x - \mu)^n f(x) dx \tag{2.7}$$

where $\mu = E(X)$ is the population mean and $f(x)$ is the probability density function (as defined in Section 1.3.2). This leads to a number of useful quantities such as

$$\mu_1 = 0 \tag{2.8}$$

$$\mu_2 = E\left[(X - \mu)^2\right] = V[X] = \sigma^2 \tag{2.9}$$

$$\mu_3 = E\left[(X - \mu)^3\right] = \sigma^3 \gamma_1 \tag{2.10}$$

$$\mu_4 = E\left[(X - \mu)^4\right] = \sigma^4 \gamma_2 \tag{2.11}$$

where γ_1 is the *skewness* (degree of asymmetry) and γ_2 is the *kurtosis* (degree of peakedness). Skewness is a measure of symmetry or, more precisely, the lack of symmetry (Davis, 2002). A distribution or dataset is symmetrical if it looks similar on either side of the "center." Kurtosis is a measure of whether the data are heavy-tailed or light-tailed relative to a normal distribution (i.e., a symmetrical distribution around the mean whose characteristics are further defined in Chapter 3). In other words, datasets with high kurtosis tend to have heavy tails or outliers. Datasets with low kurtosis tend to have light tails or the lack of outliers. Skewness and kurtosis are useful tools in classical statistics for determining if (and how) a variable should be transformed into a normal distribution for subsequent analysis, but they are not widely used in petroleum geoscience applications.

2.1.4 Graphing Univariate Data

A common approach for displaying univariate data involves a box-and-whisker plot (generally referred to as a box or Tukey plot). The "box" in the box plot shows the range between the first quartile (i.e., that value, *below* which lies 25% of the samples) and the third quartile (i.e., that value, *above* which lies 25% of the samples). The solid line within the box shows the location of the median. The ends of the whiskers connect the box to either (a) minimum and maximum of the sampled data, (b) 5th and 95th percentiles (with the outliers shown as individual symbols), or (c) other custom choices.

A companion to the box plot is the *bean plot*, which displays more information regarding the relative frequency of different values in the dataset. Each bean consists of a density trace (i.e., a smoothed estimate of the empirical probability density function corresponding to the data), which is mirrored to form the polygon shape resembling a bean. Inside the bean, a scatterplot shows the individual values as one small line for each observation.

Fig. 2.3 shows a comparison of box and bean plots for the recovery efficiency from enhanced oil recovery projects (expressed as a percentage of the original oil in place (OOIP)) from three different sources (BOX_BEAN_FIG2-3.DAT). Compared to the box plot, the bean

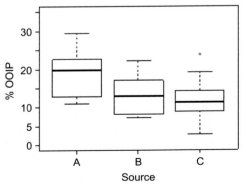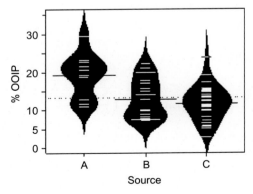

FIG. 2.3   **Example box and bean plots.**

plot is much more useful in showing how the data from sources A, B, and C are bimodal, left-skewed, and symmetrical, respectively.

2.2 BIVARIATE DATA

In this case, our interest is in describing the relationship between two variables. These bivariate measures are described below, along with some common graphical methods for visually examining and summarizing the data.

2.2.1 Covariance

The *covariance* or joint variance between two random variables is an extension of the concept of variance and is defined as

$$\text{Cov}[XY] = \sigma_{xy} = E\left[(X - \bar{X})(Y - \bar{Y})\right] = \frac{1}{N-1}\sum_{i=1}^{N}(x_i - \bar{X})(y_i - \bar{Y})$$

$$= \frac{N}{N-1}\left\{E[XY] - E[X]E[Y]\right\} \tag{2.12a}$$

The covariance can be thought of as generalization of variance. For example, if we consider the covariance of a variable with itself,

$$\text{Cov}[XX] = \sigma_{xx} = E\left[(X - \bar{X})(X - \bar{X})\right] = \text{Var}[X] \tag{2.12b}$$

However, note that the variance will always be positive, whereas the covariance can be positive or negative.

2.2.2 Correlation and Rank Correlation

The *correlation coefficient* (CC) between two random variables is a measure of the strength of their linear relationship. It is closely linked to the concept of covariance and is defined as

$$CC = \rho_{xy} = \frac{\sigma_{xy}}{\sigma_x \sigma_y} = \frac{1}{N-1}\sum_{i=1}^{N}\left(\frac{x_i - \bar{X}}{\sigma_x}\right)\left(\frac{y_i - \bar{Y}}{\sigma_y}\right) \tag{2.13}$$

The value of CC ranges between −1 (indicating perfectly negative correlation) and +1 (indicating perfectly positive correlation). The sign indicates the direction of the trend (i.e., positive or negative), and the absolute value quantifies the strength of the relationship. It is important to note that the concept of correlation strictly applies for a monotonic relationship.

If the variables of interest are related in a nonlinear manner, then the *rank correlation coefficient* (*RCC*) can be used as a more robust measure of (nonlinear) association. It is computed by calculating the correlation coefficient between the ranks of the original variables. Here, rank transformation implies assigning *rank* = 1 to the smallest value, *rank* = 2 to the next highest value, and so on. This is the simplest nonparametric linearizing technique that does not require assuming any functional form for the relationship (Iman and Conover, 1983). The computation of *RCC* is based on the following:

$$RCC = \rho_{xy(rank)} = \frac{\sigma_{xy(rank)}}{\sigma_{x(rank)}\sigma_{y(rank)}} = \frac{1}{N-1}\sum_{i=1}^{N}\left(\frac{R_{x,i}-\bar{R}_x}{\sigma_{R_x}}\right)\left(\frac{R_{y,i}-\bar{R}_y}{\sigma_{R_y}}\right) \tag{2.14}$$

A simpler alternative for calculating *RCC* is based on the difference of ranks, *d*, namely,

$$RCC = 1 - \frac{6\sum d^2}{N(N^2-1)} \tag{2.15}$$

Note also that *CC* is also referred to as the *Pearson* correlation coefficient, whereas *RCC* is referred to as the *Spearman* correlation coefficient.

Table 2.3 shows an example calculation of correlation and rank correlation. Here, ϕ is porosity, *K* is permeability, $R(\phi)$ is the rank of porosity, $R(K)$ is the rank of permeability, and *d* is the absolute difference between the two sets of ranks. Note that the first set of calculations of $\rho[\phi K]$ and $\rho[R\_\phi K]$ is based on Eqs. (2.12)–(2.14), whereas the second calculation of $\rho[R\_\phi K]$ is based on Eq. (2.15). The prefix "*R*" denotes rank transformed. In general, the Pearson correlation coefficient will be much more sensitive to data clusters and outliers compared with the Spearman correlation coefficient. So, it is often desirable to compute both the measures to examine the robustness of the correlation.

TABLE 2.3 **Example Calculation of Correlation and Rank Correlation**

ϕ	K	ϕK	$R(\phi)$	$R(k)$	$R\_\phi K$	d
0.1	25	2.5	1	2	2	1
0.2	17	3.4	2	1	2	1
0.3	42	12.6	3	4	12	1
0.4	41	16.4	4	3	12	1
0.5	65	32.5	5	5	25	0
$E[\phi]$	$E[K]$	$E[\phi K]$	$SD[\phi]$	$SD[k]$	$Cov[\phi K]$	$\rho[\phi K]$
0.3	38	13.5	0.158	18.5	2.6	0.890
$E[R\_\phi]$	$E[R\_K]$	$E[R\_\phi K]$	$SD[R\_\phi]$	$SD[R\_k]$	$Cov[R\_\phi K]$	$\rho[R\_\phi K]$
3	3	10.6	1.581	1.6	2	0.8

$\rho[\phi K]$ = 2.6/0.158/185.15 = 0.890

$\rho[R\_\phi K]$ = 2/1.581/1.6 = 0.8

$\rho[R\_\phi K]$ = 1 − (6 * (1^2 + 1^2 + 1^2 + 1^2)/5/(5^2 − 1)) = 0.8

2.2.3 Graphing Bivariate Data

A *scatterplot* between two variables is the simplest way of graphically displaying their relationship. The strength of linear association, if any, is given by the absolute value of the Pearson CC, ρ, whereas the sign of ρ indicates whether the correlation is positive or negative.

Multiple example scatter diagrams are shown in Fig. 2.4, displaying a range of possible behavior (SCATTER_FIG2.4.DAT) between two generic variables, X and Y. In the top-left panel, (A) a strong positive linear trend can be discerned, which corresponds to $\rho = 0.734$. The top-right panel (B) indicates a very strong negative linear trend, corresponding to $\rho = -0.893$. The bottom-left panel (C) shows a weak negative correlation corresponding to $\rho = -0.145$, while a modest positive correlation with $\rho = 0.484$ is displayed in the bottom-right panel (D). In general, the value of ρ is inversely proportional to the degree of scatter around the underlying linear trend (shown as the dashed lines in Fig. 2.4).

Scatterplots can also be used to demonstrate the utility of rank transformation. Consider the porosity data given earlier in Table 2.1, where the corresponding permeability values k (mD) are [12, 30, 62, 6.7, 5.7, 14, 2.6, 3, 33, 8, 40, 23, 20, 3.1, 1.2, 110, 100, 84, 58, 38, 27] (PERM_FIG2-5.DAT). Fig. 2.5A shows the porosity-permeability scatterplot for this dataset, indicating an apparent exponential relationship. On the other hand, Fig. 2.5B shows the same data after rank transformation, where a much stronger linear trend can be observed. This is consistent with the Pearson CC value of 0.789 for these data, which reflects the strength of the linear trend in Fig. 2.5A, and the Spearman CC value of 0.916 reflecting the strength of the rank-transformed linear trend in Fig. 2.5B.

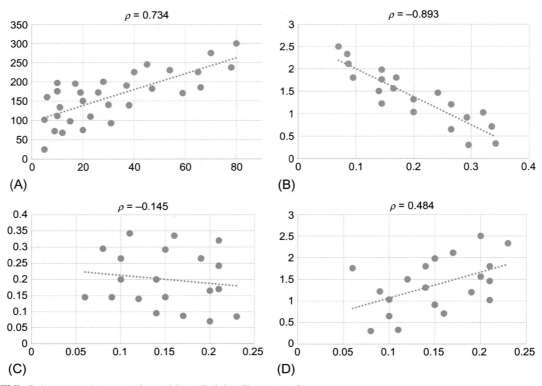

FIG. 2.4   **Example scatterplots with underlying linear trend.**

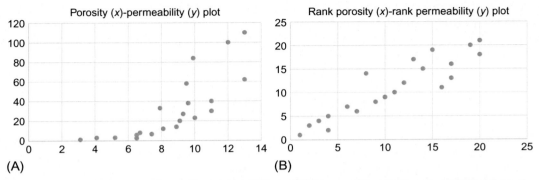

FIG. 2.5   Scatterplots showing (A) nonlinear relationship and (B) improved linearization after rank transformation.

It should be noted that the Pearson CC between porosity and natural log of permeability is calculated to be 0.922, that is, essentially the same as the Spearman CC. This confirms the power of the rank transformation to linearize data in a nonparametric manner without making any assumptions about the functional form of the underlying relationship.

Another important point is that the square of the Pearson CC is the same as the coefficient of determination (R^2) of linear regression, as will be shown later in Chapter 4. This means that the standard goodness-of-fit measure for a linear relationship can be directly determined from the Pearson CC without going through the regression process.

As shown in Fig. 2.6, scatterplots can also be combined with histograms, that is, bar charts that display how the individual variables are distributed within their respective ranges

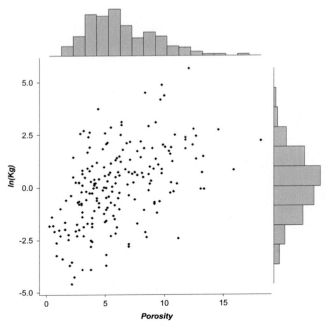

FIG. 2.6   Example scatterplot showing histograms of the individual variables.

(SCATTER_FIG2-6.DAT). These histograms, shown along the axes, represent the marginal (individual) distributions of X and Y, whereas the scatterplot is a representation of the joint distribution between X and Y. More will be said in Chapter 3 about the computation and mathematical representation of histograms.

2.3 MULTIVARIATE DATA

The analysis of correlation in multivariate data is a simple extension of the concepts discussed previously for bivariate data. This involves calculating the Pearson (or Spearman) CC for all variable pairs and presenting it in the form of a *correlation matrix*. Table 2.4 shows an example for a three-variable dataset (COR_TAB2-4.DAT). Since the correlation matrix is symmetrical, it is sufficient to show only the lower (or upper) part of the matrix.

Similarly, the concept of scatterplots for data visualization can be generalized to a *scatterplot matrix* or a *pairs plot*, which is generated by combining scatterplots of all variable pairs to show their interrelationship (Venables and Ripley, 1997). Each scatterplot can be annotated with a smoothing line that helps visualize the underlying trend (linear or otherwise) and can also be color coded to indicate membership of individual data points in different groups. The histograms for each individual variable are sometimes presented along the diagonal. The advantage of such a plot is that an overview of the relationships, patterns, and/or trends among independent variables, as well as between dependent and independent variables, can be obtained at the same time.

An example scatterplot matrix is shown in Fig. 2.7 (PAIRS_FIG2.7.DAT) from a numerical study of CO_2 injection potential into a deep saline aquifer (Mishra et al., 2014).

TABLE 2.4   **Calculation of Correlation Matrix**

X1	X2	X3	X1	X2	X3
0.295	0.3	0.08	0.342	0.33	0.11
0.32	1.02	0.21	0.095	1.8	0.14
0.242	1.46	0.21	0.2	1.03	0.1
0.14	1.5	0.12	0.2	1.31	0.14
0.265	0.65	0.1	0.087	2.11	0.17
0.335	0.71	0.16	0.145	1.22	0.09
0.085	2.33	0.23	0.145	1.76	0.06
0.17	1.8	0.21	0.165	1.56	0.2
0.265	1.2	0.19	0.145	1.98	0.15
0.292	0.91	0.15	0.07	2.5	0.2
		X1	**X2**	**X3**	
	X1	1			
	X2	−0.89255	1		
	X3	−0.14474	0.484331	1	

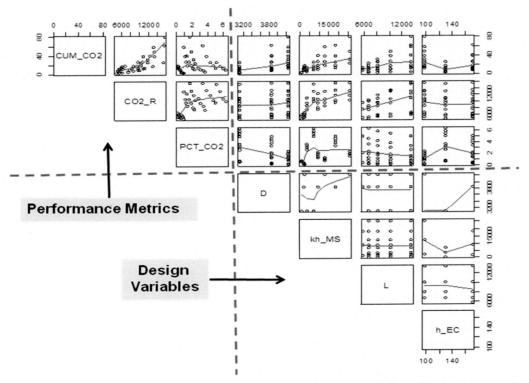

FIG. 2.7 **Example scatterplot matrix.** *Data from Mishra, S., Oruganti, Y., Sminchak, J., 2014. Parametric analysis of CO₂ sequestration in closed volumes. Environ. Geosci. 21(2), 59–74.*

Here, CUM_CO2 is the total CO_2 injected over 30 years, CO2_R is the radius of the CO_2 plume, PCT_CO2 is the percent dissolved in the aqueous phase, D is the depth to the injection zone, kh_MS is the permeability-thickness production of the target reservoir (Mount Simon sandstone formation), L is the well spacing, and h_EC is the thickness of the caprock (Eau Claire shale). Note that the first three variables along the diagonal (i.e., performance metrics) are the simulation results, whereas the other four (i.e., design variables) are the model inputs.

It should be noted that some of the scatterplots show nonmonotonic relationships (e.g., between CUM_CO2 and h_EC), which points to the limitation of using linear correlation to describe the strength of association. The Pearson CC for this case is close to zero, indicating no apparent linear correlation between the two variables. However, there is clearly a quadratic-type nonrandom relationship, as indicated by the smoothing red line. For such cases, the concept of *mutual information* has been proposed as a more robust measure of nonrandom association that can handle both monotonic and nonmonotonic relationships (Mishra et al., 2009). This is further discussed in section 6.4.4.

2.4 SUMMARY

In this chapter, we have discussed a number of measures for describing and visualizing univariate, bivariate, and multivariate data. These include measures of central tendency and spread, as well as correlation and rank correlation. The concepts are explained through several worked problems.

Exercises

1. The following table shows average values of porosity and permeability from multiple core samples collected at three different wells. Calculate the following: (a) $E[\phi_{avg}]$, $SD[\phi_{avg}]$, $E[k_{avg}]$, and $\rho[\phi_{avg}, k_{avg}]$.

Well	No. of Samples	Average ϕ	Average k
1	45	0.24	41
2	27	0.32	65
3	62	0.19	17

2. Using the data given in Table 2.1 [POR_TAB2-1.DAT], create 3 independent subsets and calculate the arithmetic, geometric, and harmonic mean porosity for each 7-sample subset. Compare them to the full 21-sample result.
3. Calculate the variance for each of the 7-sample subsets in problem 2, and compare the 7-sample subsets to the full 21-sample result. How does the result change if you combine the 7-sample subsets into 14-sample subsets?
4. Derive the result: $Cov[XX] = Var[X]$.
5. Prepare box and bean plots for the data in Table 2.1 [POR_TAB2-1.DAT].
6. Let X = permeability, Y = porosity, Z = connate water saturation. The means of X, Y, and Z are 20 mD, 0.15, and 0.30, respectively. (a) Assuming that permeability and connate water saturations are uncorrelated, find $E[XZ]$. (b) If $E[XY] = 10$ and $E[YZ] = 0.2$, find the covariance between (X,Y) and (Y,Z).
7. Find the raw and rank correlation coefficient for ϕ and $\ln(k)$ using the data given in Table 2.3. Comment on their similarity (or lack thereof).
8. For the data shown in Fig. 2.6 [SCATTER_FIG2-6.DAT], create subsets consisting of the first 10, 50, and 100 samples. Calculate the RCC for each subset, and compare with the RCC for the full dataset. Comment on the possible reasons for differences, if any.
9. Prepare a scatter plot matrix for the data given in Table 2.4 [COR_TAB2-4.DAT].

References

Davis, J.C., 2002. Statistics and Data Analysis in Geology. John Wiley & Sons, New York, NY.
Iman, R.L., Conover, W.J., 1983. A Modern Approach to Statistics. John Wiley and Sons, New York, NY.
Jensen, J., Lake, L.W., Corbett, P., Goggin, D., 2000. Statistics for Petroleum Engineers and Geoscientists. Elsevier, New York, NY.

King, M.J., MacDonald, D.G., Todd, S.P., Leung, H., 1998. Application of Novel Upscaling Approaches to the Magnus and Andrew Reservoirs. Society of Petroleum Engineers, Richardson, TX. https://doi.org/10.2118/50643-MS.

Mishra, S., Deeds, N.E., Ruskauff, G.J., 2009. Review paper – global sensitivity analysis techniques for groundwater models. Ground Water 47 (5), 730–747.

Mishra, S., Oruganti, Y., Sminchak, J., 2014. Parametric analysis of CO_2 sequestration in closed volumes. Environ. Geosci. 21 (2), 59–74.

Venables, W.N., Ripley, B.D., 1997. Modern Applied Statistics With S-PLUS, second ed. Springer, New York, NY.

Distributions and Models Thereof

The topic of this chapter is probability distributions, which help us describe and visualize data. We discuss methods for describing empirical data, as well as theoretical models that can be used as mathematical representations of distributions.

3.1 EMPIRICAL DISTRIBUTIONS

Distributions are a means of expressing uncertainty in data in terms of the range of possible values and their likelihood. Sampled data are generally represented empirically in terms of frequency plots (histograms) and/or cumulative probability (quantile) plots. The probability of any outcome is inferred from the observed frequency over a long sequence of trials (as opposed to being based on subjective judgment).

3.1.1 Histogram

The histogram is an empirical (sampled) form of the probability density function (PDF), which characterizes the theoretical frequency of occurrence corresponding to a given interval. It is constructed by first dividing the observed range into several intervals (bins) and plotting the actual frequency of occurrence in each interval.

The number of bins used in histograms is usually a matter of trial and error. Common rules of thumb that have been proposed include the following:

- For a sample size of N, the number of intervals k should be the smallest integer such that $2^k \geq N$ (Iman and Conover, 1983).
- A default value for the number of bins is $\{3.3\log(N) + 1\}$, which is only a suggestion and is often exceeded (Venables and Ripley, 1997).

Because the shape of the histogram is strongly dependent on the number of intervals chosen, it is not a very robust graphic tool. As an example, Fig. 3.1 shows the histograms

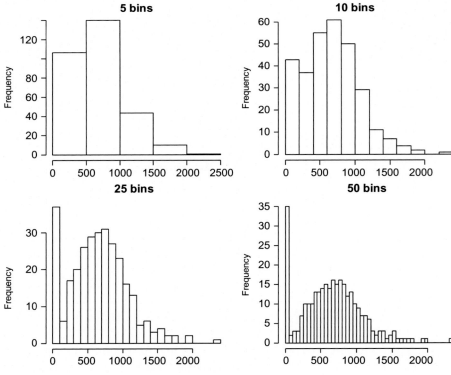

FIG. 3.1   **Histograms showing sensitivity to bin size.**

corresponding to 5, 10, 25, and 50 bins generated for observed wind speed data using a sample size of 300. The bimodal character of the data (i.e., a high proportion of very low values) is only evident in histograms with 25 bins or higher. Both of the binning rules cited above suggest using 10 bins or lower. Thus, it is always useful for the analyst to experiment with multiple bin sizes until a robust indication of the shape of the PDF is obtained.

3.1.2 Quantile Plot

The quantile plot is an empirical (sampled) form of the cumulative distribution function (CDF), which characterizes the probability that the observed value of a random variable is smaller than some specified value. To construct a quantile plot, the data are first ranked in ascending order from the smallest (x_1) to the largest (x_N), where N is the number of samples. For each sorted value, x_i, the quantile (cumulative frequency) is determined as $q_i = i/(N+1)$, and the quantile plot is generated by plotting q_i versus x_i. Percentiles are obtained by multiplying the quantile values by 100. The quantile plot is also referred to as an empirical CDF.

Compared with the histogram, the quantile plot is a much more robust tool for visualizing the fraction of samples that fall below a given value and for determining if a distribution is symmetrical or skewed. Some useful diagnostic rules for symmetry evaluation are listed below:

- A symmetrical distribution is characterized by an S-shaped quantile plot, where the distance on the horizontal axis between the median (50th percentile) and any percentile P below the median is equal to the distance from the median to the (100-Pth percentile). Symmetrical distributions are characterized by mean = median = mode.
- If the distribution has positive skewness, that portion of the quantile plot corresponding to $q > 0.9$ will usually be longer and flatter than the rest of the plot.
- Conversely, distributions with negative skewness have a long flat portion on the quantile plot corresponding to $q < 0.1$.

Examples of these characteristics are presented in Fig. 3.2 using well-log data from the Salt Creek field (SALT-CREEK.DAT) discussed further in Sections 4.4 and 5.4. The top panel shows the empirical CDF and histogram corresponding to a symmetrical distribution using data for the logarithm of microspherically focused log (MSFL). The middle panel shows a distribution with negative skewness using data for the bulk density log (RHOB). Finally, the bottom panel shows a distribution with positive skewness using data for the gamma ray log (GR).

3.2 PARAMETRIC MODELS

Parametric models of continuous probability distributions (i.e., mathematical relationships given in terms of one or more parameters) are useful for several reasons:

- They provide a compact mathematical construct for summarizing empirical data.
- They allow extrapolation of data beyond the observed minimum and maximum values and interpolation between sampled data points.

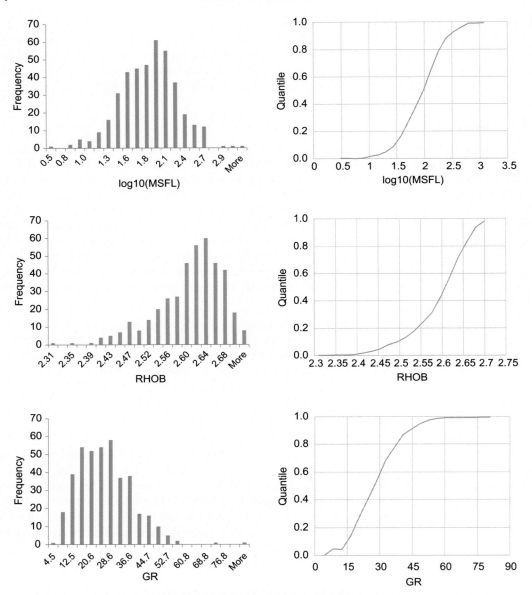

FIG. 3.2   Characteristic shapes of quantile plots and their corresponding histograms for a symmetrical distribution *(top)*, negatively skewed distribution *(middle)*, and positively skewed distribution *(bottom)*.

- They enable the statistical representation of uncertain quantities based on purely physical or mechanistic considerations.
- They facilitate the Bayesian updating of distributions based on prior information.

Some of the common parametric models useful for petroleum geoscience applications are described below. Here, $f(x)$ denotes the PDF, $F(x)$ denotes the CDF, μ denotes the mean, and σ denotes the standard deviation for the theoretical distribution assigned to the random variable of interest, X. This discussion is based on standard references dealing with statistical applications in engineering and geoscience (e.g., Ang and Tang, 1975; Harr, 1987; Morgan and Henrion, 1990; Jensen et al., 2000; Davis, 2002).

3.2.1 Uniform Distribution

The uniform distribution is useful as a rough model for representing low states of knowledge when only the upper and lower bounds are known. All possible values within the specified maximum and minimum values are equally likely:

$$\text{PDF}: f(x) = \frac{1}{b-a} \; ; a \le x \le b \tag{3.1}$$

where $b = $ maximum and $a = $ minimum:

$$\text{CDF}: F(x) = \frac{x-a}{b-a} \tag{3.2}$$

$$\text{Moments}: \quad \mu = \frac{(a+b)}{2} \; ; \sigma^2 = \frac{(b-a)^2}{12} \tag{3.3}$$

Notation: $X \sim U(a, b)$

The log-uniform distribution is a variation of the uniform, where the inputs cover a large range (e.g., multiple orders of magnitude), but nothing else is known about the shape of the underlying distribution. If x is such an uncertain quantity of interest, then $\ln(x)$ is taken to be uniformly distributed. Fig. 3.3 shows a schematic of the uniform distribution.

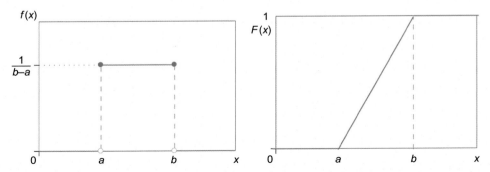

FIG. 3.3 **Schematic of a uniform distribution showing PDF** *(left)* **and CDF** *(right)*. *Source: https://commons.wikimedia.org/w/index.php?curid=27378784.*

EXAMPLE 3.1 Working with uniform distributions

The datum pressure in a reservoir has a mean of 2800 psi and a standard deviation of 100 psi based on data from a number of new wells. Under the assumption that the data follow a uniform distribution, calculate the 10th and 90th percentile values (i.e., P10 and P90) for this distribution.

Solution

It can be shown that the parameters of the uniform distribution can be expressed in terms of the sample moments as follows:

$$a = \mu - \sqrt{3}\sigma \; ; b = \mu + \sqrt{3}\sigma$$

Thus, $a = 2800 - \sqrt{3}*100 = 2626.8$ psi.
And $b = 2800 + \sqrt{3}*100 = 2973.2$ psi.

From Eq. (3.2), we have for the P10 value
$0.1 = (x - 2626.8)/(2973.2 - 2626.8)$
Therefore, $x = 2661.4$ psi (at P10).

Similarly, for the P90 value, we have
$0.9 = (x - 2626.8)/(2973.2 - 2626.8)$
Therefore, $x = 2938.6$ psi (at P90).

3.2.2 Triangular Distribution

The triangular distribution can be used as an improvement over the uniform distribution for modeling situations where nonextremal (central) values are more likely than the upper or lower bounds. It is useful as a rough model when minimum, maximum, and most likely values are known—typically on the basis of subjective judgment:

$$\text{PDF}: \; f(x) = \begin{cases} \dfrac{2(x-a)}{(b-a)(c-a)} & ; a \leq x \leq c \\ \dfrac{2(b-x)}{(b-a)(b-c)} & ; c < x \leq b \end{cases} \tag{3.4}$$

where b = maximum, a = minimum, and c = mode (most likely value):

$$\text{CDF}: \; F(x) = \begin{cases} \dfrac{(x-a)^2}{(b-a)(c-a)} & ; a \leq x \leq c \\ 1 - \dfrac{(b-x)^2}{(b-a)(b-c)} & ; c < x \leq b \end{cases} \tag{3.5}$$

$$\text{Moments}: \; \mu = \frac{(a+b+c)}{3} \; ; \sigma^2 = \frac{(a^2 + b^2 + c^2 - ab - bc - ca)}{18} \tag{3.6}$$

Notation: $X \sim T(a, b, c)$

Depending on the location of the modal value, triangular distributions can be symmetrical or asymmetrical. When uncertainties are large and asymmetrical and/or the range between

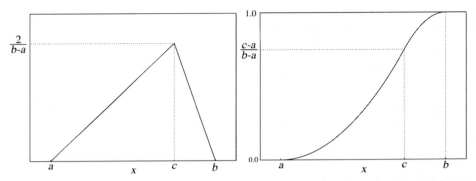

FIG. 3.4   **Schematic of a triangular distribution showing PDF** *(left)* **and CDF** *(right). Source: https://commons. wikimedia.org/w/index.php?curid=182089.*

the minimum and maximum spans several orders of magnitude, a log-triangular distribution may be more appropriate. Fig. 3.4 shows a schematic of the triangular distribution. Note that the probability density reaches a maximum at the mode, c, and is equal to $2/(b-a)$, with the corresponding cumulative probability being equal to $(c-a)/(b-a)$.

EXAMPLE 3.2 Working with triangular distributions

Well-log-derived water saturations in an oil field appear to follow a triangular distribution with minimum = 17%, mode = 28%, and maximum = 49%. Calculate the 10th and 90th percentile values (i.e., P10 and P90) for this distribution and the cumulative probability corresponding to the mode.

Solution

Given, $a = 0.17$, $c = 0.28$, and $b = 0.49$

We first calculate the cumulative probability corresponding to the mode,

$P_{mode} = (c-a)/(b-a) = (0.28-0.17)/(0.49-0.17)$

$P_{mode} = 0.34$

Since P10 is to the left of the mode, we use the first expression in Eq. (3.5) leading to

$0.1 = (x-0.17)^2/((0.49-0.17)(0.28-0.17))$

Therefore, $x = 0.229$ (at P10).

For the P90 value, we use the second expression in Eq. (3.5) leading to

$0.9 = 1-(0.49-x)^2/((0.49-0.17)(0.49-0.28))$

Therefore, $x = 0.408$ (at P90).

3.2.3 Normal Distribution

The normal distribution is the commonly used "bell curve" for modeling unbiased uncertainties and random errors of the additive kind and symmetrical distributions of many natural processes and phenomena. A commonly cited rationale for assuming normal distributions is the central limit theorem (cf. Section 3.4), which states that the sum of independent observations asymptotically approaches a normal distribution regardless of the shape of the underlying distribution(s):

$$\text{PDF}: \; f(x) = \frac{1}{\sqrt{2\pi\sigma^2}} \exp\left\{ -\frac{1}{2}\left(\frac{x-\mu}{\sigma}\right)^2 \right\}; \; -\infty \leq x \leq \infty \tag{3.7}$$

where μ = mean and σ = standard deviation.

CDF: $F(x)$ has no closed-form solution but is often presented using the complementary error function solution. However, it can also be expressed in terms of the standard normal CDF, $G(\cdot)$, tabulated in many statistics texts and available as the intrinsic function NORMSINV in Microsoft Excel:

$$F(x) = G\left(\frac{x-\mu}{\sigma}\right) \tag{3.8}$$

Moments: Same as parameters of the distribution:

Notation: $X \sim N(\mu, \sigma)$

The symmetrical nature of the distribution is often characterized in terms of the probability coverage corresponding to a given interval around the mean. For example, the interval $[\mu \pm 1\sigma]$ corresponds to $P = 0.683$, the interval $[\mu \pm 2\sigma]$ corresponds to $P = 0.954$, and the interval $[\mu \pm 3\sigma]$ corresponds to $P = 0.997$—as shown in Fig. 3.5. This implies that in a normal distribution, about 68% of the samples fall between $[\mu \pm 1\sigma]$, about 95% fall between $[\mu \pm 2\sigma]$, and virtually all samples fall between $[\mu \pm 3\sigma]$. These properties are used to assign confidence intervals to estimates as discussed later.

The normal distribution is often used as a "default" distribution for representing uncertainties. Because the distribution is theoretically unbounded, care should be taken to ensure that the standard deviation is not so large as to result in nonphysical sampled values at the lower tail. Fig. 3.6 shows examples of normal distributions. Additional details regarding the normal distribution are discussed in Section 3.3.

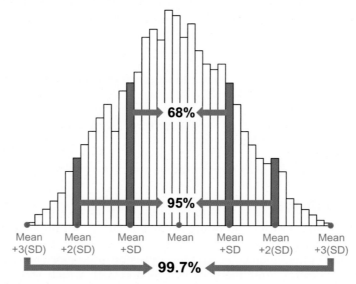

FIG. 3.5   **Schematic of normal distribution showing probability coverages.**

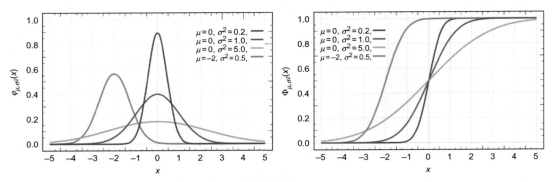

FIG. 3.6 **Examples of normal distributions showing PDF** *(left)* **and CDF** *(right). Source: https://commons. wikimedia.org/w/index.php?curid=3817954.*

3.2.4 Lognormal Distribution

The lognormal distribution is widely used for representing skewed, nonnegative, physical quantities. It is useful as an asymmetrical model for multiplicative independent uncertainties. As with the normal distribution, the rationale for assuming a lognormal distribution is based on the central limit theorem, which states that the product of independent observations asymptotically approaches a lognormal distribution—regardless of the shape of the underlying distribution(s):

$$\text{PDF}: \quad f(x) = \frac{1}{x\sqrt{2\pi\beta^2}} \exp\left\{ -\frac{1}{2}\left(\frac{\ln(x) - \alpha}{\beta} \right)^2 \right\} \; ; 0 \leq x \leq \infty \tag{3.9}$$

where α = mean of $\ln(x)$ and β = standard deviation of $\ln(x)$.

CDF: $F(x)$ has no closed-form solution. However, it can be expressed in terms of the standard normal CDF, $G(\cdot)$, tabulated in many statistics texts and available as the intrinsic function NORMSDIST in Microsoft Excel:

$$F(x) = G\left(\frac{\ln(x) - \alpha}{\beta} \right) \tag{3.10}$$

$$\text{Moments}: \quad \mu = \exp\left(\alpha + \frac{\beta^2}{2} \right) \; ; \sigma^2 = \mu^2\left\{ \exp\left(\beta^2\right) - 1 \right\} = \exp\left(2\alpha + 2\beta^2\right) \tag{3.11}$$

Notation: $X \sim LN(\alpha, \beta)$

Here, the geometric mean or median is given by e^α, while the quantity e^β is referred to as the geometric standard deviation. Fig. 3.7 shows an example of a lognormal distribution.

A commonly used measure of spread, especially for lognormal permeability distributions, is the Dykstra-Parsons coefficient (Willhite, 1986), V_{DP}, defined as

$$V_{DP} = \left(k_{50} - k_{84.1}\right)/k_{50} \tag{3.12}$$

where k_P is the Pth percentile value as obtained from a lognormal fit to the sampled distribution of permeability, k. It is readily shown that $V_{DP} = 1 - \exp(-\beta)$ with β as defined above (Mishra et al., 1991). Fig. 3.8 illustrates the graphic calculation of the Dykstra-Parsons coefficient. Additional details regarding the lognormal distribution are discussed in Section 3.3.

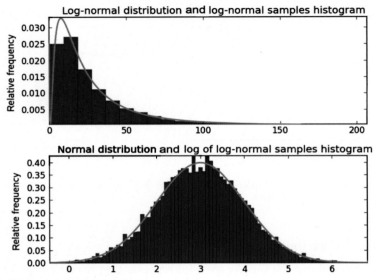

FIG. 3.7   **Example of a lognormal distribution.**

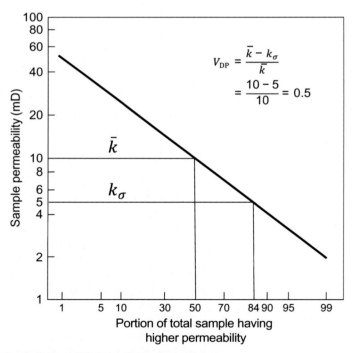

FIG. 3.8   **Graphical calculation of Dykstra-Parson's coefficient.**

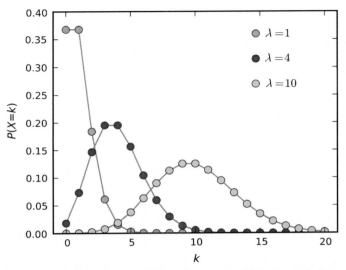

FIG. 3.9   **Examples of Poisson distribution.** *Source: https://commons.wikimedia.org/w/index.php?curid=9447142.*

3.2.5 Poisson Distribution

When events occur as a purely random (Poisson) process, the number of independent events occurring within a fixed time interval follows a Poisson distribution. The number of events is discrete and constrained to nonnegative integers:

$$\text{PDF}: \quad f(x) = \frac{\alpha^x \exp(-\alpha)}{x!} \; ; x = 0, 1, 2, 3, \dots. \tag{3.13}$$

where $\alpha =$ parameter of the distribution:

$$\text{CDF}: \quad F(x) = \sum_{k=0}^{x} \frac{\alpha^x \exp(-\alpha)}{x!} \tag{3.14}$$

$$\text{Moments}: \quad \mu = \alpha; \sigma^2 = \alpha \tag{3.15}$$

$$\text{Notation}: \quad X \sim Po(\alpha)$$

The Poisson distribution can be used to model such quantities as the number of earthquakes happening during a given period and the number of lost days from equipment failure in a year. Fig. 3.9 shows an example Poisson distribution.

EXAMPLE 3.3 Working with Poisson distributions

Records collected from a downhole gauge from an offshore platform over a 4-year period show that there have been 24 days during which the signal was completely or partially lost. What is the probability that there would be zero lost signal days next year? What is the probability that the signal would be lost for 10 days over the year?

Solution

We assume that the Poisson distribution is an appropriate model for this dataset with $\alpha = \lambda t$ where $\lambda =$ mean occurrence rate and t is time.

From the data, $\lambda = 24/4 = 6$ per year, and $t = 1 \rightarrow \alpha = 6$.

For zero occurrences of lost signal, $x = 0$ and Eq. (3.13) gives

$$f(0) = 6^0 \exp(-6)/0!; f(0) = 0.0025$$

The probability of 10 lost signal days is, with $x = 10$,

$$f(10) = 6^{10} \exp(-6)/10!; f(0) = 0.0413$$

3.2.6 Exponential Distribution

The exponential distribution is used to model the time between the occurrence of events in an interval of time (e.g., time between successive failures of a machine) or the distance between events in space (e.g., distance between successive breaks in a pipeline):

$$\text{PDF}: \quad f(x) = \lambda e^{-\lambda x} \tag{3.16}$$

where $\lambda =$ parameter of the distribution:

$$\text{CDF}: \quad F(x) = 1 - e^{-\lambda x} \tag{3.17}$$

$$\text{Moments}: \quad \mu = 1/\lambda ; \sigma^2 = 1/\lambda^2 \tag{3.18}$$

$$\text{Notation}: \quad X \sim Exp(\lambda)$$

As shown in Fig. 3.10, the Poisson and exponential distributions are closely related. If the number of occurrences in a unit interval can be represented by a Poisson distribution with parameter λ, then the time between successive occurrences (i.e., the x_i values shown below) will follow an exponential distribution also with parameter λ. In other words, if the mean number of occurrences per time interval is λ, then the mean length of time between successive occurrences is $1/\lambda$. In Example 3.3, there were six lost signal days per year. Therefore, the mean length of time between successive lost signal days is $1/6$ years, that is, 61 days.

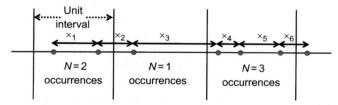

FIG. 3.10   **Relationship between Poisson and exponential distributions.**

3.2.7 Binomial Distribution

A binomial distribution is the distribution of the number of successes k in a sequence of n independent trials, where the probability of success p is constant from trial to trial. Each trial with two outcomes (success or failure) is also called a Bernoulli experiment:

$$\text{PMF}: \quad f(k; n, p) = {}^{n}C_k p^k (1-p)^{n-k} ; \; {}^{n}C_k = \frac{n!}{(n-k)!k!} \tag{3.19}$$

where k = number of successes:

$$\text{Moments}: \quad \mu = np; \; \sigma^2 = np(1-p) \tag{3.20}$$

$$\text{Notation}: \quad X \sim B(n, p)$$

The binomial distribution can be approximated by the Poisson distribution as n becomes large (i.e., >20) and p becomes small (i.e., <0.05) such that the product np is constant. Also, the binomial distribution can be approximated by the normal distribution if n is large and p approaches 0.5 such that $[np(1-p)] \geq 25$.

EXAMPLE 3.4 Working with binomial distributions

A company is planning a six-well gas exploration program in a new shale basin. Based on its experience with similar basins, the probability of success is assumed to be 10%. What is the probability that the no wells will discover gas? What is the probability of one success?

Solution

For the first part (which is sometimes referred to as the "gambler's ruin" case), $k = 0$, $n = 6$, and $p = 0.1$.

From Eq. (3.19), we have

$$f(0; 6, 0.1) = {}^{6}C_0 (0.1)^0 (1 - 0.1)^{6-0}; \; f(0; 6, 0.1) = 0.531$$

Similarly, when $k = 1$ instead, we have

$$f(1; 6, 0.1) = {}^{6}C_1 (0.1)^1 (1 - 0.1)^{6-1}; f(1; 6, 0.1) = 0.354$$

3.2.8 Weibull Distribution

The Weibull distribution is widely used to represent distributions of process performance metrics such as completion time or equipment failure rate. Because of its flexibility to assume negatively skewed, symmetrical, or positively skewed shapes, it can also be used to represent many nonnegative physical quantities:

$$\text{PDF}: \quad f(x) = \frac{k}{\lambda} \left(\frac{x}{\lambda}\right)^{k-1} \exp\left\{-\left(\frac{x}{k}\right)^k\right\} ; k, \lambda > 0, 0 \leq x \leq \infty \tag{3.21}$$

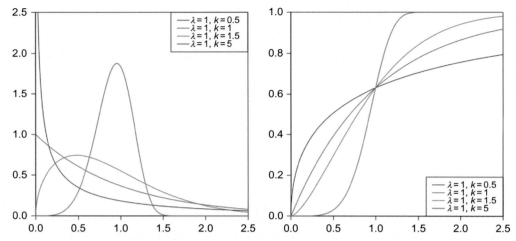

FIG. 3.11   **Examples of Weibull distributions showing PDF** *(left)* **and CDF** *(right). Source: https://commons.wikimedia.org/w/index.php?curid=9671812.*

where $\lambda =$ scale parameter and $k =$ shape parameter:

$$\text{CDF:}\quad F(x) = 1 - \exp\left\{ -\left(\frac{x}{\lambda}\right)^k \right\} \tag{3.22}$$

$$\text{Moments:}\quad \lambda = \beta\Gamma\left(1 + \frac{1}{k}\right) ;\; \sigma^2 = \lambda^2\left\{ \Gamma\left(1 + \frac{2}{k}\right) - \Gamma^2\left(1 + \frac{1}{k}\right) \right\} \tag{3.23}$$

Notation: $X \sim W(k, \lambda)$

Here, $\Gamma(.)$ is the incomplete gamma function. The scale parameter, λ, is that value of time at which the CDF is equal to 0.632 (or $1 - 1/e$). The shape parameter, k, indicates whether the process of interest (i.e., failure rate) is decreasing with time ($k < 1$), constant ($k = 1$), or increasing with time ($k > 1$). Fig. 3.11 shows some example Weibull distributions.

The Weibull distribution is a commonly used tool for modeling growth (or decline) in biological, clinical, population, and natural resource studies. It has also been used to analyze production decline from unconventional reservoirs (Mishra, 2012). This entails multiplying both the PDF (Eq. 3.21) and CDF (Eq. 3.22) by a carrying capacity, M, which denotes the physical production limit on the system and provides an upper bound on resource extraction. The CDF can be interpreted as cumulative production and the PDF as the instantaneous production rate.

EXAMPLE 3.5 Working with Weibull distributions

Production decline data from an unconventional gas reservoir were found to be fit in a Weibull model with scale parameter $\lambda = 89.5$ months and shape parameter $k = 0.765$. When will the field recover 50% of the producible gas reserves?

Solution

We are solving for
$$0.5 = F(t) = 1 - \exp(-(t/\lambda)^k) = 1 - \exp(-(t/89.5)^{0.765})$$
Thus, $t = 55.5$ months.

3.2.9 Beta Distribution

The beta distribution is a very flexible model for describing random proportions and for characterizing uncertainty over a fixed range (i.e., with finite upper and lower bounds). It can take both symmetrical and skewed shapes within the prescribed interval:

$$\text{PDF}: \quad f(x) = \frac{x^{\alpha-1}(1-x)^{\beta-1}}{B(\alpha,\beta)} \; ; \alpha,\beta > 0, 0 \leq x \leq 1 \tag{3.24}$$

where α, β = distribution parameters and $B(\alpha,\beta) = \Gamma(\alpha)\Gamma(\beta)/\Gamma(\alpha+\beta)$.

CDF: $F(x)$ has no closed-form solution but can be expressed using the intrinsic function BETADIST in Microsoft Excel:

$$\text{Moments}: \quad \mu = \frac{\alpha}{\alpha+\beta} \; ; \sigma^2 = \frac{\alpha\beta}{(\alpha+\beta)^2(\alpha+\beta+1)} \tag{3.25}$$

Notation: $X \sim Beta(\alpha,\beta)$

Fig. 3.12 shows some example beta distributions.

The beta distribution does not have a mechanistic basis but can be very useful for fitting empirical data to distributions, because of the flexible mathematical form of Eq. (3.24). This becomes particularly relevant for the purposes of uncertainty quantification using Monte Carlo simulation, as discussed in Chapter 6.

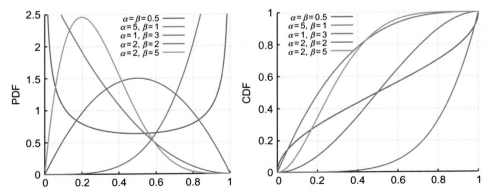

FIG. 3.12   **Examples of beta distributions showing PDF** *(left)* **and CDF** *(right). Source: https://commons. wikimedia.org/w/index.php?curid=15404569.*

3.3 WORKING WITH NORMAL AND LOG-NORMAL DISTRIBUTIONS

Since normal and lognormal distributions are commonly used to represent many variables in the petroleum geosciences, it is useful to have a better understanding of how these distributional models can be manipulated to provide the information needed by the analyst.

3.3.1 Normal Distribution

For the normal distribution, recall that the CDF, $F(x)$, has no closed-form solution and is expressed in terms of the standard normal CDF, $G(\cdot)$:

$$F(x) = G\left(\frac{x - \mu}{\sigma}\right) = G(z) \tag{3.26}$$

where $z = (x - \mu)/\sigma$ is the standard normal variate (also known as the z-score). Note that z is a dimensionless variable with zero mean and unit variance. Eq. (3.26) can be rewritten as

$$z = \frac{x - \mu}{\sigma} = G^{-1}\{F(x)\} = G^{-1}(q) \tag{3.27}$$

with the quantile, q, being used as an approximation of the cumulative probability, F. This leads to the following compact equation for representing a normal distribution:

$$x = \mu + \sigma G^{-1}(q) = \mu + \sigma z \tag{3.28}$$

Note that the inverse normal CDF or the z-score can be readily calculated using the intrinsic Microsoft Excel function, NORMSINV.

A partial table for the $G(z)$ function is shown below in Table 3.1, indicating the cumulative probability levels corresponding to different values of the standard normal variate, z. Recall that the z-value (also referred to as the normal score) defines the normalized separation from the mean (in terms of standard deviation units) for a given quantile or percentile of interest. Thus, the probability coverage corresponding to mean $\pm 3SD = 0.9987 - 0.0013 = 0.997$, mean $\pm 2SD = 0.9772 - 0.0228 = 0.954$, and mean $\pm 1SD = 0.8413 - 0.1587 = 0.683$ (as shown earlier in Fig. 3.5). From this table of values, we also note that the commonly used P10 (10th percentile) in petroleum geoscience uncertainty analysis studies corresponds to $z = -1.28$, the P50 corresponds to $z = 0$, and the P90 corresponds to $z = 1.28$.

TABLE 3.1 Tabulated Values of the Z-Function

$z = G^{-1}(q)$	−3.5	−3	−2.5	−2	−1.64	−1.5	−1.28	−1
$G(z)$	0.0002	0.0013	0.0062	0.0228	0.0505	0.0668	0.1003	0.1587
$z = G^{-1}(q)$	−0.67	−0.5	−0.25	0	0.25	0.5	0.67	1
$G(z)$	0.2514	0.3085	0.4013	0.5000	0.5987	0.6915	0.7486	0.8413
$z = G^{-1}(q)$	1.28	1.5	1.64	2	2.5	3	3.5	4
$G(z)$	0.8997	0.9332	0.9495	0.9772	0.9938	0.9987	0.9998	1.0000

Manipulation of the normal distribution can be relatively straightforward by realizing that the random variable, x, can be readily mapped onto the corresponding standard normal variate, z, if the mean, μ, and standard deviation, σ, are known. This is illustrated below for a simple example.

EXAMPLE 3.6 Working with normal distributions

Given that the thickness of a shale formation, h, is assumed to be a normally distributed variable with mean $\bar{h}=60$ ft and coefficient of variation $CV[h]=20\%$, determine (a) probability that the thickness is between 45 and 75 ft, that is, $P[45\leq h\leq 75]$ and (b) the 95th percentile value.

Solution

Standard deviation $\sigma[h]=\bar{h} * CV[h]=60 * 0.2=12$ ft.

(a) We have $z(h=75)=(75-60)/12=1.25$, and $z(h=45)=(45-60)/12=-1.25$.

By interpolating from Table 3.2 or using the intrinsic Microsoft Excel function NORMSDIST, $G(1.25)=0.894$, $G(-1.25)=1-(G(1.25))=0.106$

Thus, $P[45\leq h\leq 75]=G[z(h=75)]-G[z(h=45)]=G(1.25)-G(-1.25)=0.894-0.106$:

$$P[45\leq h\leq 75]=0.788$$

(b) From Table 3.1, we note that when $q=0.95$, $z=1.64$.

Therefore, $h_{0.95}=\bar{h}+z*\sigma[h]=60+1.64*12$:

$$h_{0.95}=79.7\ ft$$

3.3.2 Normal Score Transformation

Often, it is useful to transform a sample distribution into the space of an equivalent normal distribution, where many statistical operations can be easily performed and visualized. This is particularly true for problems in geostatistics, as many of the spatial modeling algorithms (e.g., sequential Gaussian simulation) are restricted to normally distributed random variables. The approach involves a rank-preserving one-to-one transformation, as schematically shown in Fig. 3.13. For any given value x_i of the original variable, the empirical cumulative probability or quantile q is set equal to the cumulative probability for a standard normal distribution, and the equivalent z-value or normal score is calculated. Mathematically, this can be stated as

$$z=G^{-1}(q)=G^{-1}\left(\frac{rank(x_i)}{n+1}\right) \tag{3.29}$$

Once the required mathematical operations are carried out in terms of the transformed normal score z, the results can be readily back-transformed into the space of the original variables.

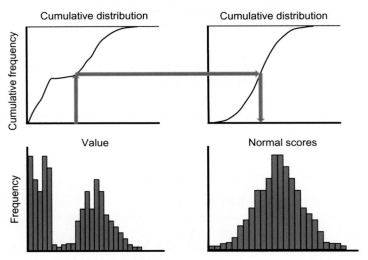

FIG. 3.13   **Schematic of normal score transformation.**

An example normal score transformation, using the bulk density (RHOB) data for the Salt Creek field discussed earlier in Fig. 3.2, is presented in Fig. 3.14. Note how the asymmetry of the original data is modified via the normal score transformation.

3.3.3 Log-Normal Distribution

For a lognormal distribution, we define the standard normal variate as

$$z = \frac{\ln(x) - \alpha}{\beta} = G^{-1}\{F(x)\} = G^{-1}(q) \tag{3.30}$$

where $\alpha =$ mean and $\beta =$ standard deviation of $\ln(x)$. Rearranging, we get

$$\ln(x) = \alpha + \beta G^{-1}(q) \tag{3.31}$$

which is a compact equation, similar to Eq. (3.28), for representing the lognormal distribution.

Essentially, working with lognormal distributions involves transforming the data into logarithmic space, determining the parameters α and β, performing the necessary operations as for the normal distribution case, and back-transforming into the space of original coordinates. The following relationships are useful in this context:

- The geometric mean, e^{α}, is also the median (i.e., $F_{0.5}$).
- The geometric standard deviation, e^{β}, is the ratio $F_{0.84}/F_{0.5}$ or $F_{0.5}/F_{0.16}$.
- The nature of the distribution is fundamentally multiplicative (and therefore additive after logarithmic transformation).

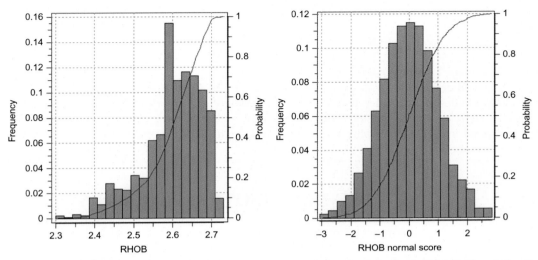

FIG. 3.14 **Histogram and CDF of the bulk density log data for the Salt Creek field** *(left)* **and for the corresponding normal score transform** *(right).*

EXAMPLE 3.7 Working with lognormal distributions

Permeability values from a reservoir are known to follow a lognormal distribution with $\alpha = 3.61$ and $\beta = 0.67$. Calculate the geometric mean (GM), geometric standard deviation (GSD), arithmetic mean (AM) and standard deviation (ASD), P10 and P90 values, and the Dykstra-Parsons coefficient for this distribution.

Solution

1. Geometric mean $= \exp(\alpha) = \exp(3.61)$; GM $= 37$.
2. Geometric standard deviation $= \exp(\beta) = \exp(0.67)$; GSD $= 1.95$.

   Also, since the 84th and 16th percentile values are one standard deviation away from the mean for a normally distributed variable, we can write

   $$k_{0.84} = \exp(\alpha + \beta) = \exp(3.61 + 0.67) = 72.2$$

   $$k_{0.16} = \exp(\alpha - \beta) = \exp(3.61 - 0.67) = 18.9$$

   Therefore, geometric standard deviation $= k_{0.84}/k_{0.5} = 72.2/37$; GSD $= 1.95$
   Alternatively, geometric standard deviation $= k_{0.5}/k_{0.16} = 37/18.9$; GSD $= 1.96$
   Thus, all calculations are essentially consistent.

3. Arithmetic mean $= \exp(\alpha + \beta^2/2) = \exp(3.61 + 0.67^2/2)$; $AM = 46.3$.

   Arithmetic standard deviation $= \exp(2\alpha + 2\beta^2) = \exp(2*3.61 + 2*0.67^2)$; $ASD = 57.9$.

4. From Table 3.1, the P10 and P90 values correspond to $z = -1.28$ and 1.28, respectively.

   Therefore, $k(P10) = k_{0.1} = \exp(\alpha - 1.28\beta) = \exp(3.61 - 1.28*0.67)$; $k_{0.1} = 15.7$.
   Also, $k(P90) = k_{0.9} = \exp(\alpha + 1.28\beta) = \exp(3.61 + 1.28*0.67)$; $k_{0.9} = 87.1$.

5. The Dykstra-Parsons coefficient, V_{DP}, can be calculated in two different ways as follows:

$$V_{DP} = (k_{50} - k_{16.1})/k_{50} = (37 - 18.9)/37\,;\ V_{DP} = 0.49$$

Also, $V_{DP} = 1 - \exp(-\beta) = 1 - \exp(-0.67);\ V_{DP} = 0.49.$

Note that definition of V_{DP} given earlier in Section 3.2.4 uses a notation where permeability is arranged in descending order, that is, higher permeability values correspond to lower probability levels. Here, we have used the more conventional notation where higher permeability values correspond to higher probability levels, which require a slight modification to the V_{DP} formula as shown above.

3.4 FITTING DISTRIBUTIONS TO DATA

Although many theoretical distributions can be used to fit an empirical dataset, only a handful of distributions are considered in practice. The key features of these are described in Table 3.2.

Some issues worth considering during the selection of a probability distribution include

- physical or mechanistic basis for choosing a distribution family and/or shape;
- discrete versus continuous nature of variable;
- physical bounds, if any, for the variable;
- nature and degree of skewness, if any;
- importance of extreme values (tails).

With many geologic variables, the choice of an appropriate physical or mechanistic basis for assigning a probability distribution is often difficult. This is particularly true when the variable of interest represents behavior spanning several geologic regimes (e.g., depositional environments), or the variable is sought to be characterized over a scale that is different from the scale of observations. In these situations, a graphic analysis of the data using special

TABLE 3.2　**Commonly Used Distribution Models**

Distribution	Useful for Representing
Uniform (log-uniform) Triangular (log-triangular)	Low state of knowledge and/or subjective judgment
Normal	Errors due to additive processes
Lognormal	Errors due to multiplicative processes
Poisson	Frequency of rare events
Exponential	Occurrence times of random event
Binomial	Number of successes in a sequence of trials
Weibull	Component failure rates
Beta	Bounded, unimodal, random variables

probability plots can help identify candidate distributions and/or eliminate inappropriate parametric models. Once a distribution has been selected, its parameters can then be estimated using one of several techniques discussed below. Also, statistical goodness-of-fit tests can be applied to further refine and/or validate the choice of distributions. This sequence of (a) hypothesizing a family of distributions, (b) estimating distribution parameters, and (c) assessing quality of fit of parameters is described below, along with illustrative examples for some commonly used distributions.

3.4.1 Probability Plots

Probability plots are useful for comparing the distribution of empirical data to postulated theoretical distributions. The observations are plotted, generally after some transformation, so that they would fall approximately on a straight line if the assumed parametric model was the "true" distribution from which the observations were sampled. Given that deviations from a straight line can be readily identified, probability plotting provides a straightforward screening tool for distribution selection (D'Agostino and Stephens, 1986).

A visual examination of the probability plot will often help in determining whether the postulated distribution is appropriate or not. The analyst should also apply his or her knowledge of the process and/or parameter to verify that the agreement between the observations and the theoretical distribution is acceptable in key data regimes (e.g., high/low values). In mentally weighting portions of the data differently, the analyst should be aware of deviations from the straight line that commonly occur at the tails due to the finite size of samples.

The starting point in probability plotting is an empirical CDF or quantile plot, where the quantiles (cumulative frequency) of the empirical distribution are plotted against the corresponding observations. Two common choices for defining the quantile, q, are the Weibull plotting position:

$$q_i = \frac{i}{N+1} \qquad\qquad (3.32a)$$

and the Hazen plotting position:

$$q_i = \frac{i-0.5}{N} \qquad\qquad (3.32b)$$

where i is the rank of the observation (sorted from smallest to largest) and N is the number of observations. Both of these approaches ensure that the minimum and maximum values of the sample are not assigned cumulative probabilities of 0 and 1, respectively.

A probability plot is a graph of the ranked observation, x_i, versus an approximation of the expected value of the inverse CDF, $F^{-1}(q_i)$. The relationships needed to construct probability plots are discussed below for the normal and lognormal distributions, because of their ubiquity in petroleum geoscience problems. D'Agostino and Stephens (1986) provide such relationships for several other distributions.

The basic expression for a normal probability plot follows from the linearized representation of normal distributions as discussed earlier in Section 3.3.1, namely,

$$x = \mu + \sigma G^{-1}(q) \tag{3.33a}$$

which suggests that a graph of x versus $G^{-1}(q)$, or z, should yield a straight line if the observed data follow a normal distribution. The straight line is characterized by a slope equal to the standard deviation, σ, and intercept equal to the mean, μ.

Similarly, the basic expression for a lognormal probability plot follows from the linearized representation of lognormal distributions as discussed earlier in Section 3.3.2, namely,

$$\ln(x) = \alpha + \beta G^{-1}(q) \tag{3.33b}$$

Thus, a graph of $\ln(x)$ versus $G^{-1}(q)$, or z, should yield a straight line if the observed data follow a lognormal distribution. The straight line is characterized by a slope equal to the standard deviation, β, and intercept equal to the mean, α, of the transformed variable $\ln(x)$. Note that the arithmetic mean and the arithmetic standard deviation can be readily obtained using Eq. (3.11).

3.4.2 Parameter Estimation Techniques

Once a candidate distribution has been selected for a dataset, the parameters of the postulated theoretical distribution can be obtained in a variety of ways. The easiest approach is to use linear regression in conjunction with probability plots. Additional techniques include nonlinear least-squares analysis or the method of moments—as described below.

Linear Regression Analysis

In the previous section, transformations were described for linearizing the relationship between observed (sampled) values and the corresponding quantiles of the postulated distribution. The slope and intercept of the resulting straight line in a probability plot were seen to be related to the parameters of the underlying distribution. These relationships are summarized in Table 3.3.

Note that in this approach, the estimated parameters are derived from an analysis based on a transformation of the parametric distribution to a linear form. Therefore, these parameters may not produce the most optimal fit to the distribution when transformed back to the original scale. Although more advanced techniques such as nonlinear least-squares analysis can be used to improve such estimates, the linearized approximations should provide a good first approximation, especially if the probability plot produces a good fit.

TABLE 3.3   Linearizing Relationships for Normal and Lognormal Distributions

Distribution	Y-Axis	X-Axis	Slope	Intercept
Normal	x	$G^{-1}(q)$	σ	μ
Lognormal	$\ln(x)$	$G^{-1}(q)$	β	α

Method of Moments

In the method of moments approach, the parameters of a probability distribution model are estimated by matching the moments of the dataset with that of the candidate model. The number of moments required corresponds to the number of unknown model parameters. Application of this method is straightforward, as closed-form expressions for the moments can be readily derived for most common distributions. However, the raw moments may be biased due to the presence of outliers and/or the lack of perfect agreement between the data and the model.

Equations relating the theoretical first two moments (i.e., mean and variance) to distributional parameters were presented in Section 3.2. These provide the basis for estimating the parameters of the distribution from the sample moments (identified by the "hat" symbol) computed as follows:

$$\hat{\mu} = \frac{1}{N}\sum_{i=1}^{N} x_i$$

$$\hat{\sigma}^2 = \frac{1}{N-1}\sum_{i=1}^{N}(x_i - \hat{\mu})^2 \tag{3.34}$$

Method of moment estimators for some of the common distributions are

$$\text{Normal:} \quad \mu = \hat{\mu} \; ; \sigma = \hat{\sigma} \tag{3.35a}$$

$$\text{Lognormal:} \quad \beta^2 = \ln\left(1 + \frac{\hat{\sigma}^2}{\hat{\mu}^2}\right) \; ; \alpha = \ln(\hat{\mu}) - \frac{\beta^2}{2} \tag{3.35b}$$

Nonlinear Least-Squares Analysis

A more flexible approach involves the use of nonlinear least-squares analysis, where the goal is to estimate model parameters such that the mean squared difference between the observed and predicted CDF is minimized. This process can be readily implemented using the nonlinear optimization package SOLVER in Microsoft Excel:

1. Set up the data in a two-column format, with the dependent variable being the observed quantile, q_i, and the independent variable being the observed value, x_i.
2. Compute the sample moments, $\hat{\mu}$ and $\hat{\sigma}$.
3. Estimate the parameters of the postulated model from the sample moments using Eq. (3.35a,b) or equivalent expressions from Section 3.2. These will be used as initial guesses for the nonlinear regression.
4. Calculate the theoretical cumulative probability, F_i, using the appropriate form of the postulated parametric model as given in Section 3.2 and estimates of model parameters obtained from step 3.
5. Compute the difference between F_i and q_i.
6. Set up SOLVER to minimize the sum of the squares of the differences in step 5, by adjusting the parameters estimated in step 3.

EXAMPLE 3.8 Fitting a normal distribution

Porosity values from multiple core samples in a well (POR_TAB2-1.DAT) were found to be ϕ (%) $=$(3.1, 4.1, 5.2, 6.5, 6.5, 6.7, 7.4, 7.9, 8.1, 8.9, 9.1, 9.3, 9.5, 9.6, 9.9, 10, 11, 11, 12, 13, and 13). Fit these data to a normal distribution, and calculate the parameters of the distribution.

Solution

A normal distribution was first fit to the data using the probability plotting method. This requires plotting ϕ against the inverse of the standard normal CDF, $G^{-1}(q)$, where q is the quantile. As shown in Fig. 3.15, a very good fit was obtained except at the extreme tails, with an R^2 value approximately equal to 1. The lognormal parameters are calculated from the slope and intercept of the best-fit line on the probability plot as $\mu = 8.76$ and $\sigma = 3.09$.

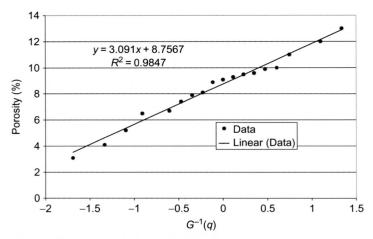

FIG. 3.15   **Probability-plot-based normal distribution fit, Example** 3.3.

Next, these parameters are obtained using nonlinear least-squares analysis, which requires minimizing the sum of the squared differences between the observed and the predicted quantiles corresponding to each observed value. The Excel function NORMSDIST was used to generate the standard normal CDF necessary for estimating the cumulative probability. The corresponding best-fit parameters, obtained using the SOLVER toolbox in Excel, are $\mu = 8.81$ and $\sigma = 2.92$, which agree very well with those estimated using the probability plotting method. Fig. 3.16 compares the observed CDF with the predictions using regression parameters.

Some of the calculational details for the distribution process are shown below in Table 3.4. The quantile values are calculated using Eq. (3.32a), and the corresponding $G^{-1}(q)$ values are obtained from the NORMSINV function. This leads to the probability plot shown in Fig. 3.14. The least-squares fitting for the CDF follows the stepwise procedure described earlier and leads to Fig. 3.15.

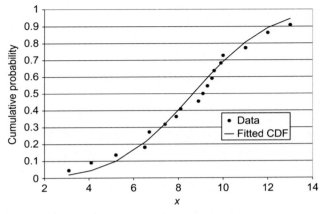

FIG. 3.16   **Nonlinear-regression-based fit to normal CDF, Example** 3.3.

TABLE 3.4   **Calculational Details for Fitting a Normal Distribution**

Rank	Porosity (%) ϕ	Quantile q	Probability Plotting $G^{-1}(q)$	x	Least-Squares Fitting $F(x)$	diff
1	3.1	0.045455	−1.69062	3.1	0.025192	−0.02026
2	4.1	0.090909	−1.33518	4.1	0.053251	−0.03766
3	5.2	0.136364	−1.0968	5.2	0.107989	−0.02837
4	6.5	0.181818	−0.90846	6.5	0.214194	0.032376
5	6.5	0.181818	−0.90846	6.5	0.214194	0.032376
6	6.7	0.272727	−0.60459	6.7	0.234705	−0.03802
7	7.4	0.318182	−0.47279	7.4	0.314321	−0.00386
8	7.9	0.363636	−0.34876	7.9	0.377385	0.013749
9	8.1	0.409091	−0.22988	8.1	0.403677	−0.00541
10	8.9	0.454545	−0.11419	8.9	0.512052	0.057507
11	9.1	0.5	0	9.1	0.539323	0.039323
12	9.3	0.545455	0.114185	9.3	0.56641	0.020956
13	9.5	0.590909	0.229884	9.5	0.593189	0.00228
14	9.6	0.636364	0.348756	9.6	0.606425	−0.02994
15	9.9	0.681818	0.472789	9.9	0.645346	−0.03647
16	10	0.727273	0.604585	10	0.658011	−0.06926
17	11	0.772727	0.747859	11	0.773256	0.000528
18	11	0.772727	0.747859	11	0.773256	0.000528
19	12	0.863636	1.096804	12	0.862623	−0.00101
20	13	0.909091	1.335178	13	0.924321	0.01523
21	13	0.909091	1.335178	13	0.924321	0.01523

The method of moments can also be used to estimate the normal parameters as per Eq. (3.35a,b). The sample mean and standard deviation are found to be $\mu = 8.66$ and $\beta = 2.69$. These values are generally consistent with those obtained from probability plotting and nonlinear least-squares analysis. The discrepancy in the estimated value of σ is likely due to the small sample size of 21. The method of moment estimates is best used as initial guesses for the nonlinear regression.

3.5 OTHER PROPERTIES OF DISTRIBUTIONS AND THEIR EVALUATION

3.5.1 Central Limit Theorem and Confidence Limits

Consider a sequence of random variables, $x_1, x_2, \ldots, x_n$, that are independent and identically distributed with mean $= \mu$ and standard deviation $= \sigma$. Then, the sample mean $\overline{X}_n$ also becomes a random variable, with mean and variance given by

$$E\left[\overline{X}_n\right] = \mu \tag{3.36}$$

$$V\left[\overline{X}_n\right] = \frac{\sigma^2}{n} \tag{3.37}$$

This is known as the *law of large numbers* and holds regardless of the shape of the underlying distribution. It is important to note here that while the population mean is a constant quantity, the sample mean is a random variable. The standard deviation of the mean is also referred to as the standard error (s_e).

Furthermore, a well-known result from statistics invokes the *central limit theorem* to state that if n is reasonably large, no matter what the distribution of x, the sample mean will be approximately normal (Davis, 2002) with mean and standard deviation as given above. In other words, the sum of independent observations will asymptotically approach a normal distribution. By extension, the product of independent observations will asymptotically approach a lognormal distribution. This asymptotic normality property holds regardless of the shape of the underlying distributions, as demonstrated in Fig. 3.17 for three different distributions—uniform (symmetrical), exponential (moderately skewed), and lognormal (strongly skewed). The convergence to normality increases with sample size for all three distributions. However, the greater the skewness, the larger the number of samples needed for the sampling mean to be approximately normal.

This information is useful in assigning confidence intervals (i.e., error bars) that take into account the dependence on sample size whenever population parameters are approximated by sample estimators. For example, the $100(1-\alpha)\%$ confidence interval for sample mean (when the population variance is known or assumed to be the same as the sample variance) is given by

$$CI = \left(\overline{X}_n \pm z_{\alpha/2} \frac{\sigma}{\sqrt{n}}\right) \tag{3.38}$$

where Z is the standard normal distribution $N(0,1)$ and $z_{0.025} = 1.96$ (≈ 2) when $\alpha = 0.05$. Note that a significance level of $\alpha = 0.05$ implies that there is a 1 in 20 chances that the result could be

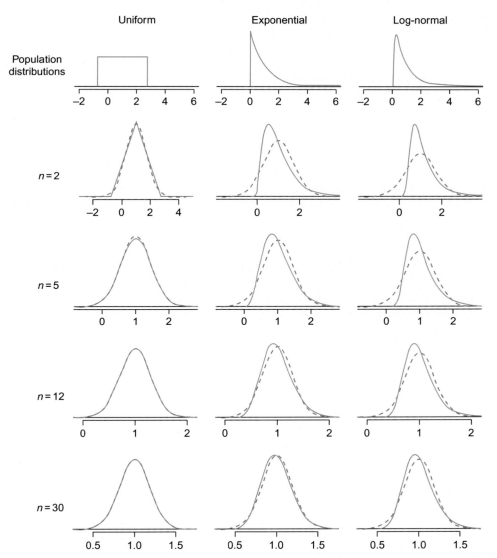

FIG. 3.17 **Sampling distribution of the mean for three distributions at various sample sizes.** *From Boundless. "Examining the Central Limit Theorem." OpenIntro Statistics Boundless, 20 Sep. 2016. Retrieved Jun. 8, 2017 from https://www.boundless.com/users/233402/textbooks/openintro-statistics/foundations-for-inference-4/examining-the-central-limit-theorem-36/examining-the-central-limit-theorem-176-13789/*

incorrect. Although significance levels of $\alpha = 0.05$ or $\alpha = 0.01$ (i.e., 1 in 100 chances of error) have been commonly used in experimental statistics, these are arbitrary thresholds, and it is incumbent upon the analyst to determine the problem-specific acceptable level of error when seeking a decision from a statistical test.

When the sample size n is large, σ can be replaced by sample standard deviation(s). For small sample sizes (typically < 30), the sample estimate s can show considerable variability, and $\overline{X}_n$ may not be normally distributed. However, if the population is approximately normal, we can still use the quantity $\frac{\overline{X}_n - \mu}{s/\sqrt{n}}$, but it is no longer normally distributed. Instead, it follows student's t distribution with $(n-1)$ degrees of freedom (Davis, 2002).

Specifically, if $x_1, x_2, x_3 \cdots, x_n$ is a set of random values from the population that has a normal distribution, the standardized variable t is defined as follows:

$$t = \frac{\overline{X}_n - \mu}{s/\sqrt{n}} \tag{3.39}$$

The spread of t curves is affected by the number of degree of freedom ($\nu = n - 1$). As the degree of freedom increases, t curves approach z curves, the standard normal distribution (Fig. 3.18).

Returning to the problem of confidence intervals for small sample sizes, the $100(1\alpha)\%$ confidence interval for sample mean (when the population variance is unknown) is given by

$$CI = \left(\overline{X}_n \pm t_{n-1,\alpha/2} \frac{s}{\sqrt{n}} \right) = \left(\overline{X}_n \pm t_{n-1,\alpha/2} s_e \right) \tag{3.40}$$

where s^2 is the sample variance and s_e is the standard error of the mean. As shown earlier, the t distribution not only is quite similar to the standard normal distribution in terms of its shape but also depends on the sample size. Some useful values for the t-statistic corresponding to the commonly used value of $\alpha = 0.05$ are given in Table 3.5. The t-statistic can also be evaluated using the intrinsic Microsoft Excel functions T.INV and T.DIST.

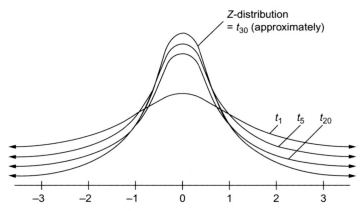

FIG. 3.18   Comparison of t distribution to standard normal (z) distribution.

TABLE 3.5   Selected Critical Values for the t-Statistic for $\alpha = 0.05$

ν	1	2	3	4	5	6	7	8	9	10
t	12.7062	4.3027	3.1824	2.7764	2.5706	2.4469	2.3646	2.306	2.2622	2.2281
ν	12	14	16	18	20	22	24	26	28	30
t	2.1788	2.1448	2.1199	2.1099	2.086	2.0739	2.0639	2.0555	2.0484	2.0423
ν	40	50	60	70	80	90	100	110	130	150
t	2.0211	2.0086	2.0003	1.9944	1.9901	1.9867	1.984	1.9818	1.9784	1.9759

EXAMPLE 3.9 Calculating confidence interval for the mean

Consider the 10-sample net pay data: h (ft), (13, 17, 15, 23, 27, 29, 18, 27, 20, and 24). Calculate the 95% confidence intervals associated with the population mean.

Solution

From the data, we have the following:

- Sample mean $\overline{X}_n = 21.3$, and sample standard deviation $s = 5.52$.
- Number of samples $n = 10$.
- Standard error $s_e = s/\sqrt{n} = 5.52/\sqrt{10} = 1.75$.
- When $n = 10$ and $\alpha = 0.05$, $t_{9,0.025} = 2.26$.
- $CI = [21.3 \pm 2.26*1.75]$; $CI_t = 17.35, 25.25$.
- If the population and sample standard deviations are assumed to be the same, the Z-distribution applies, and we get $CI = [21.3 \pm 1.96*1.75]$; $CI_z = 17.88, 24.72$.

3.5.2 Bootstrap Sampling

Often, the sample statistics of interest may include quantities other than the mean, such as the standard deviation or some tail percentile (e.g., $q_{0.95}$). Simple expressions for computing the confidence intervals of such quantities are generally not available, and one has to resort to a simulation-based approach for their estimation. One such popular technique is the bootstrap (Efron and Tibisharini, 1993). It can be described as a numerical procedure for simulating the sampling distribution of any statistic and estimating its mean, standard deviation, and associated confidence intervals. Bootstrap has become a frequently used tool in computational statistics.

Given a data of sample size n, the general approach in bootstrap simulation is to (1) assume a distribution that describes the quantity of interest; (2) perform r replications of the dataset by randomly drawing, with replacement, n values; and (3) calculate r values of the statistic of interest. Three common options for step (1) above include (a) resampling the actual dataset itself, (b) sampling the empirical CDF using a piecewise linear approximation, and (c) fitting parametric models (e.g., normal and lognormal) to the data. After step (3), confidence intervals for the quantity of interest can be obtained from the r values that approximate the sampling distribution.

EXAMPLE 3.10 Estimating the sampling distribution with the bootstrap

For the same 10-sample dataset used in Example 3.9, calculate the mean, standard deviation, and 95% confidence intervals for the sample mean.

Solution

A 200-sample bootstrap simulation was done by resampling, with replacement, the 10 originally sampled values. Shown below are the original values (in the first row), along with the first three resampled datasets. Fig. 3.19 shows a histogram of the bootstrap distribution of the mean:

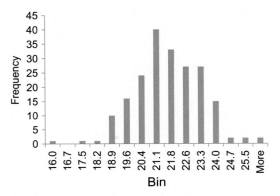

FIG. 3.19   **Bootstrap derived distribution for sample mean, Example 3.3.**

The bootstrap calculated statistics for the sample mean are (a) mean $= 21.35$, (b) standard deviation $= 1.60$, and (c) confidence interval $= 18.50, 24.73$.

3.5.3 Comparing Two Distributions

Often, it is necessary to ascertain if the mean values of two different samples are the same. This is one way of testing for the similarity between two empirical distributions. An alternative, and more comprehensive, way is to test for the similarity in the full PDF/CDF. This approach is also useful for comparing an empirical PDF/CDF against a theoretical PDF/CDF (corresponding to a postulated distribution) as a way of determining the goodness of fit. Both of these methods are described below. However, a starting point for comparing two distributions is often a graphic comparison of their quantiles, known as a quantile-quantile plot or Q-Q plot.

Q-Q Plot

The Q-Q plot refers to a graph where two distributions are compared by plotting their corresponding quantiles. Whereas histograms and summary statistics reveal gross differences, the Q-Q plot can reveal subtle differences between the distributions. A Q-Q plot of two identical distributions will be a straight line with unit slope (i.e., $x = y$). If the Q-Q plot plots as a straight line with a nonunit slope, then the two distributions have the same shape but their location and spread may differ. If the Q-Q plot displays nonlinear behavior, then the two distributions are different. Fig. 3.20 shows the Q-Q plot for the gamma ray (GR) and bulk density (RHOB) log data from the Salt Creek field (discussed earlier in Fig. 3.2). The differences in the skewness of the data are reflected in the Q-Q plot.

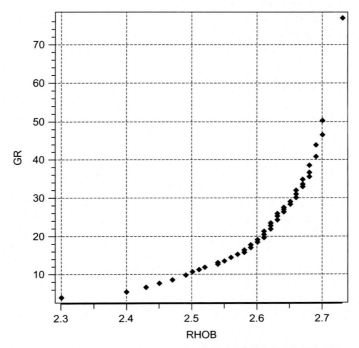

FIG. 3.20   **Q-Q plot based on the gamma ray (GR) and bulk density (RHOB) logs for the Salt Creek field data.**

Testing for Difference in Mean

The difference in the mean values obtained from samples of two different sizes can be tested for statistical significance. The difference can be small, but significant, if the sample size is large. Conversely, the difference can be large, but not significant, if the sample size is small. A quantity that measures the significance of the difference of means is based on the standard error (i.e., sample standard deviation divided by the square root of the sample size). The standard error measures the accuracy with which the sample mean estimates the "true" mean. In this approach, the t-statistic is computed based on the standard error of the difference of the means (Davis, 2002), and the significance of the difference is evaluated at an appropriate level of significance (e.g., 5%).

Let the two distributions have sample means M_1 and M_2, sample standard deviations s_1 and s_2, and sample sizes n and m. When the samples are assumed to have the same variance, the difference in means follows a t distribution, with the t-statistic calculated as

$$t = \left(\frac{M_1 - M_2}{s_e} \right) \tag{3.41}$$

where the standard error s_e can be expressed in terms of the pooled variance s_p^2 as follows:

$$s_e = s_p \sqrt{\frac{1}{n} + \frac{1}{m}} \; ; \; s_p^2 = \frac{(n-1)s_1^2 + (m-1)s_2^2}{n+m-2} \tag{3.42}$$

The t-statistic is compared with a critical value $t_{critical}$ for the degrees of freedom and the level of confidence, that is, $t_{n+m-2,\alpha/2}$. The two mean values M_1 and M_2 (and hence the two distributions) are taken to be statistically different only if the calculated t-statistic value is greater than $t_{critical}$.

If the samples are assumed to have unequal variances, then the standard error s_e is expressed in terms of the nonpooled variance as follows:

$$s_e = \sqrt{\frac{s_1^2}{n} + \frac{s_2^2}{m}} \tag{3.43}$$

and the rest of the calculational steps remain the same.

EXAMPLE 3.11 Testing for significance of difference in means

The lab-derived solution GOR for a field is 275 SCF/bbl. Production data from 10 wells indicate a range of GOR values with mean 295 and standard deviation 33.6 SCF/bbl. Is there a statistically significant increase in GOR (at the 5% significance level) indicating free gas flow has begun?

Solution

Assuming the lab measurements to have minimal error, we have
$s_e = s_e(field) = 33.6/\sqrt{10} = 10.6$ SCF/bbl
$M_1 = 295$ SCF/bbl and $M_2 = 275$ SCF/bbl
$t = (M_1 - M_2)/s_e = (295 - 275)/10.6 = 1.882$
Given $\alpha = 0.05$ and $n = 10$, $t_{critical} = t_{9,0.025} = 2.262$.
Since $t < t_{critical}$, there is no statistically significant difference between the field and lab GOR values, and free gas flow cannot be said to have begun.

Testing for Difference in Distributions

Two commonly used tests for evaluating the difference between two distributions are (a) chi-square test for binned data and (b) Kolmogorov-Smirnov test for continuous data (Davis, 2002). Key features of these two approaches are briefly described as follows.

In the *chi-square test*, the data are discretized into bins of equal probability, and the number of observations within each bin is compared with the number of expected data points. If N_i is the number of samples observed in the ith bin and n_i is the number expected according to some known distribution, then the chi-square statistic is given by

$$\chi^2 = \sum_i \frac{(N_i - n_i)^2}{n_i} \; ; i = 1, \ldots, k \tag{3.44}$$

This statistic is compared with tabulated values of the chi-square distribution for a specified confidence level with $(k-1)$ degrees of freedom. If χ^2 is larger than the tabulated critical

TABLE 3.6   Selected Critical Values for the χ^2 Statistic for $\alpha = 0.05$ and Degrees of Freedom, ν

ν	1	2	3	4	5	6	7	8	9	10
χ^2	3.84	5.99	7.81	9.49	11.07	12.59	14.07	15.51	16.92	18.31
ν	12	14	16	18	20	22	24	26	28	30
χ^2	21.03	23.68	26.3	28.87	31.41	33.92	36.42	38.89	41.34	43.77
ν	35	40	50	60	70	80	90	100	110	120
χ^2	49.77	55.76	67.5	79.08	90.53	101.88	113.15	124.34	135.48	146.57

value for the chosen level of significance and the appropriate degrees of freedom (e.g., Table 3.6), then the distributions cannot be accepted as similar.

EXAMPLE 3.12 Comparing two different distributions with the chi-square test

The number of man-hours lost by a wireline logging team because of equipment problems is given below (in row 2). Is there any significant difference between the performance of various shifts?

Solution

Given the lack of any other information, the expected distribution of equipment failures can be assumed to be uniform. Then, the analysis table (rows 3–5) can be set up as follows:

Tour	Daylight	Evening	Morning	Total
Man-hours lost	60	72	63	225
Expected	75	75	75	225
Deviation	−15	−3	+18	0
χ^2	3	0.12	4.32	7.44

From Table 3.6, $\chi^2 = 7.44$ is larger than the critical value for two degrees of freedom at the 5% significance level (i.e., 5.99), which indicates that the three groups are indeed different.

The *Kolmogorov-Smirnov test* involves a comparison between two CDFs. Both can be empirical CDFs, or one can be the theoretical CDF of a postulated distribution for the empirical data. The metric used for testing is the maximum value of the absolute difference between the two CDFs, $P(x)$ and $Q(x)$, as shown below in Fig. 3.21:

$$D = \max_x |P(x) - Q(x)| \tag{3.45}$$

The calculated value of D is compared with the tabulated value of the test statistic (as shown in Table 3.7) for the selected level of significance and the number of samples.

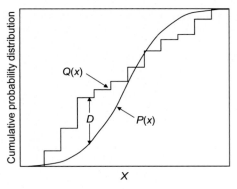

FIG. 3.21   **Schematic of Kolmogorov-Smirnov test.**

TABLE 3.7   **Selected Critical Values for the K-S Statistic for $\alpha = 0.05$**

n	1	2	3	4	5	6	7	8	9	10
D	0.995	0.7592	0.6389	0.5627	0.5088	0.4681	0.436	0.4097	0.3877	0.3689
n	12	14	16	18	20	22	24	26	28	30
D	0.3385	0.3145	0.2951	0.2789	0.2651	0.2532	0.2428	0.2335	0.2253	0.2179
n	35	40	45	50	55	60	70	80	90	100
D	0.2021	0.1894	0.1788	0.1698	0.1621	0.1553	0.144	0.1348	0.1272	0.1208

EXAMPLE 3.13 Comparing two different distributions with the K-S test

Fit a uniform distribution to the porosity data used in Example 3.2, and use the K-S test to determine if this is a statistically significant representation of the data.

Solution

The sample moments were previously determined to be $\mu=8.66$ and $\beta=2.69$. From this, we can calculate the parameters of a uniform distribution as $a=4$ and $b=13.32$ (see Example 3.1). The theoretical CDF can be calculated using Eq. (3.10) and compared with the empirical CDF. This is shown below, along with the difference:

ϕ	q	$F(x)$	Diff
3.1	0.045455	−0.09642	−0.14188
4.1	0.090909	0.010902	−0.08001
5.2	0.136364	0.12896	−0.0074
6.5	0.181818	0.268484	0.086665
6.5	0.181818	0.268484	0.086665
6.7	0.272727	0.289949	0.017221
7.4	0.318182	0.365076	0.046895

7.9	0.363636	0.418739	0.055103
8.1	0.409091	0.440204	0.031113
8.9	0.454545	0.526065	0.071519
9.1	0.5	0.54753	0.04753
9.3	0.545455	0.568995	0.02354
9.5	0.590909	0.59046	−0.00045
9.6	0.636364	0.601193	−0.03517
9.9	0.681818	0.63339	−0.04843
10	0.727273	0.644123	−0.08315
11	0.772727	0.751448	−0.02128
11	0.772727	0.751448	−0.02128
12	0.863636	0.858774	−0.00486
13	0.909091	0.966099	0.057008
13	0.909091	0.966099	0.057008

From this table, we note that $D = \max(abs(diff)) = 0.142$.

For a significance level of $\alpha = 0.95$ and a sample size of 21, the critical value of D is 0.259. Since our test statistic is below this value, we cannot conclude that the data are not drawn from a uniform distribution.

Note, however, that we would have arrived at the opposite conclusion had the sample size been ~ 75 or greater. This underscores the general uncertainty when fitting distributions to small samples, and hence, the possibility that multiple parametric models can be statistically acceptable representations of the data. If this is the case, then the model that has the best goodness-of-fit statistics (e.g., lowest RMSE) can be chosen for further analysis.

Other Methods for Comparing Distributions

Several other statistical tests can be used for comparing two distributions. These include the following:

- F-test for equality of variances
- Mann-Whitney test for equivalence of medians
- Kruskal-Wallis test for equivalence of several samples
- Wilcoxon rank-sum test for equivalence of two distributions

See Davis (2002) for additional details on these tests.

3.6 SUMMARY

In this chapter, we started with the concepts of histogram and quantile plots for summarizing empirical data. Next, we discussed a number of parametric models for describing data such as uniform, triangular, normal, log-normal, Poisson, exponential, binomial, Weibull, and beta. This was followed by methods for fitting these distributions to sample data. Finally, concepts such as central limit theorem, confidence limits, bootstrap sampling, and how to compare two distributions were presented and explained using worked problems.

Exercises

1. For the data plotted in Fig. 3.2, create new variables by taking the reciprocals of $\log_{10}(\text{MSFL})$, RHOB, and GR. Calculate and plot the corresponding histograms and the quantile plots. What do these plots tell you about the distribution of the inverse of a function?

2. For a uniform distribution, show that the parameters of the distribution are related to the sample moments as follows: $a = \mu - \sqrt{3}\sigma$, and $b = \mu + \sqrt{3}\sigma$.

3. The pore diameters (R) measured from a core sample suggest a bimodal distribution composed of two nonoverlapping log-triangular distributions, R1 and R2. The parameters of these individual distributions (in nanometers) are: for $R1 - a = 10$, $b = 30$, $c = 200$; and for R2, $a = 200$, $b = 2000$, $c = 10,000$. What is the mean and standard deviation of $\log_{10}(R)$? (Hint—write the equation for PDF as four separate terms, then use Eq. (2.7) and integrate.)

4. The number of successes in an exploratory drilling campaign is assumed to be a normal distribution with mean $= 12.5$ and standard deviation $= 3.31$. Calculate the probability of: (a) greater than 20 successes, and (b) less than 10 successes.

5. The expected value of a given rock property, η, is 30. Evaluate the probability that a random sample of this material will have a η value between 20 and 40, given $CV[\eta] = 0.12$, and assuming that the underlying distribution is (a) uniform, (b) symmetric triangular, and (c) normal.

6. Specify which of the following statements is true or false (justify your answer):

    (a) The random variable $X \sim N(5, 1)$ cannot take negative values.
    (b) For $X \sim N(8, 2)$ 68% of the X values belong to interval $6 \leq X \leq 10$.
    (c) For $\ln X \sim N(6, 2)$ 95% of the X values belong to interval $e^2 \leq X \leq e^{10}$.
    (d) For $X \sim N(7, 2)$ 16% of the data are greater than 5.

7. Over a 100-year period, the frequency of hurricanes in an offshore area was found to follow a Poisson distribution. If 13 severe hurricanes occurred during this period, calculate the probability of (a) two severe hurricanes occurring within a 3-year period, and (b) no severe hurricanes occurring within a 10-year period.

8. The failure behavior of a downhole equipment is given by a Weibull distribution with scale parameter 2.8 years and shape factor 1.6. When will 90% of the equipment need to be replaced?

9. Given the following permeability values: k (mD) $= \{40.5, 49.5, 70, 90, 110, 141, 182, 245, 405\}$, (a) fit a log-normal distribution to the data using probability plotting and calculate the geometric parameters α and β, (b) estimate the 95% confidence interval around the geometric mean.

10. Generate 1000 random variables from a uniform distribution between 0 and 1. Plot the histogram for this dataset. Divide the data into $K = 10$ columns each with $N = 100$ data values in them. For each column, compute the mean and variance (this should result in $K = 100$ mean and variance values). Plot the histogram of the mean values. What does the distribution look like? What are the mean and variance for this distribution? Is this consistent with the Central Limit Theorem? Repeat this exercise with $K = 100$ columns and $N = 10$ data values. How do the results change?

11. You are required to measure the average viscosity of a crude sample to within $\pm 3\%$. From an earlier study, you know that viscosity follows a normal distribution with mean 50 and standard deviation 5. How many measurements do you need to take?

12. Determine the 95% confidence interval for the population mean from a sample of 25 data points with sample mean $= 30$ and sample standard deviation $= 3$. How would your answer change if you ignored the fact that the standard deviation was estimated from the data?

13. Given an 18-sample permeability dataset where $K_{0.16} = 165$ mD and $K_{0.84} = 500$ mD, determine 90% confidence intervals for the geometric mean permeability. Assume K to be log-normally distributed.

14. For the dataset used in Example 3.8, generate 1000 bootstrap samples by resampling the data. Calculate and plot the histogram for the bootstrapped mean. How does this compare to the estimate using central theorem (based on sample moments)? Also, calculate and plot the histogram for the bootstrapped 10th and 90th percentile. What do these distributions look like?

15. Can we conclude (with 95% confidence level) that a set of 5 sandstone cores came from a parent population having an average porosity of 18% and a standard deviation of 5%, if the porosity of these samples was 13, 17, 15, 23, and 27 (%)?

16. Given two datasets of net pay thickness, where $\overline{X}_1 = 50$ (ft), $\overline{X}_2 = 48$ (ft), $s_1^2 = 5$, $s_2^2 = 3$, $n_1 = 25$, $n_2 = 30$, can we conclude that the two datasets have the same mean at the 95% confidence level?

17. For the dataset used in Example 3.8, fit a Weibull distribution using the approach described in Section 3.4.2. Compare the Weibull fit to the normal and uniform fits described in the chapter, using the K-S statistic as the goodness-of-fit metric.

References

Ang, A.H.-S., Tang, W.H., 1975. Probability Concepts in Engineering Planning and Design. John Wiley and Sons, New York, NY.

D'Agostino, R.B., Stephens, M.A. (Eds.), 1986. Goodness-of-Fit Techniques. Marcel Dekker, New York, NY.

Davis, J.C., 2002. Statistics and Data Analysis in Geology. John Wiley & Sons, New York, NY.

Efron, B., Tibisharini, R., 1993. An Introduction to the Bootstrap. Chapman and Hall, New York, NY.

Harr, M.E., 1987. Reliability-Based Design in Civil Engineering. McGraw-Hill, New York, NY.

Iman, R.L., Conover, W.J., 1983. A Modern Approach to Statistics. John Wiley and Sons, New York, NY.

Jensen, J., Lake, L.W., Corbett, P., Goggin, D., 2000. Statistics for Petroleum Engineers and Geoscientists. Elsevier, New York, NY.

Mishra, S., 2012. A new approach to reserves estimation in shale gas reservoirs using multiple decline curve analysis models. Society of Petroleum Engineers.https://doi.org/10.2118/161092-MS.

Mishra, S., Brigham, W.E., Orr Jr., F.M., 1991. Tracer and pressure test analysis for characterization of areally heterogeneous reservoirs. SPE Form. Eval. 6 (1), 45–54.

Morgan, M.G., Henrion, M., 1990. Uncertainty: A Guide to Dealing with Uncertainty in Quantitative Risk and Policy Analysis. Cambridge University Press, New York, NY.

Venables, W.N., Ripley, B.D., 1997. Modern Applied Statistics with S-PLUS, second ed. Springer, New York.

Willhite, G.P., 1986. Waterflooding. Society of Petroleum Engineers, Richardson, TX.

Regression Modeling and Analysis

Regression modeling is one of the most widely used tools for exploring and exploiting the relationship between dependent (response) and independent (predictor) variables. When the relationship can be expressed using linear equations (i.e., straight lines and their generalizations in multiple dimensions), it is called a linear regression. In this chapter, we start with simple linear regression involving a single predictor and response variable. We analyze the regression model in terms of residuals, variable selection, and confidence intervals for the model parameters and for model forecasts. We then generalize the concepts to multiple regression involving more than one predictor variable and nonparametric regression that derive functional relationship using flexible data-driven methods.

4.1 INTRODUCTION

In the petroleum geosciences, a very broad class of problems can be addressed using linear regression and its variations. These include permeability predictions from well logs (Wendt et al., 1986; Datta-Gupta et al., 1999), well connectivity and flow pattern analysis (Albertoni and Lake, 2003), well performance modeling (Voneiff et al., 2013), and production data analysis (LaFollette et al., 2014). The variations of linear regression often involve simple transformation of the response and/or predictor variables (e.g., logarithmic) to linearize their relationship. With little loss of generality, regression modeling concepts for continuous data can also be applied to categorical data such as geologic facies. Furthermore, using arbitrary smooth functions (*scatterplot smoothers*) for data transformation, we can extend the regression modeling to identify inherent nonlinear relationship between response and predictor variables. The generalized linear models (GLM) and alternating conditional expectation (ACE) are examples of such generalization using data transformations.

The linear parametric regression methods are model driven in the sense that they require prior knowledge of the functional relationship between the response and predictor variables. Oftentimes, such functional relationship is not readily available or difficult to ascertain, particularly in subsurface geoscience applications. As an alternative, data-driven nonparametric methods such as GLM and ACE are becoming increasingly common for petroleum reservoir characterization and analyzing data from unconventional reservoirs. We introduce the readers to GLM and ACE and conclude the chapter with a field application of ACE involving permeability predictions using well logs in a carbonate reservoir.

4.2 SIMPLE LINEAR REGRESSION

4.2.1 Formulating and Solving the Linear Regression Problem

Given the data, $(X_1, Y_1), (X_2, Y_2), \ldots \ldots (X_n, Y_n)$, we consider here a model of the form

$$Y = a + bX + \varepsilon, \ \varepsilon \sim \left(0, \sigma^2\right) \tag{4.1}$$

where a and b are parameters of regression and ε is random error that includes both measurement error and model error. If the model is adequately described by the data, the errors are expected to be independent with mean zero, $E(\varepsilon) = 0$, and a constant variance, σ^2. Fig. 4.1 shows the basic concepts in linear regression in terms of the mean values $\overline{X}$ and $\overline{Y}$, the observed values X_i and Y_i, and the predicted value $\hat{Y}_i$.

How to estimate the regression coefficients a and b? Referring to Fig. 4.1, one intuitive criterion would be to minimize the deviation between the observed value Y and predicted value $\hat{Y}$. This is commonly accomplished via the least-squares method where the goal is to minimize the residual sum of squares:

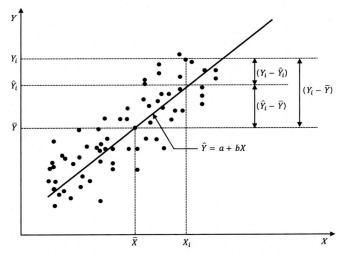

FIG. 4.1   **A linear regression model and associated quantities for model analysis.**

$$\min\, S\left(\hat{a}, \hat{b}\right) = \sum_{i=1}^{n}(Y_i - \hat{Y}_i)^2 \tag{4.2}$$

where $\hat{Y}_i = \hat{a} + \hat{b}X_i$

The minimum can be obtained as usual by taking the partial derivative of the function S with respect to estimated parameters and setting the resulting equations to zero. This procedure leads to the following solution for the least-squares regression parameters (Haan, 1986):

$$\hat{a} = \overline{Y} - \hat{b}\overline{X}$$
$$\hat{b} = \frac{S_{XY}}{S_{XX}} \tag{4.3a}$$

where

$$\overline{X} = \frac{1}{N}\sum_{i=1}^{n}X_i;\ \overline{Y} = \frac{1}{N}\sum_{i=1}^{n}Y_i \tag{4.3b}$$

$$S_{XX} = \sum_{i=1}^{n}(X_i - \overline{X})^2;\ S_{XY} = \sum_{i=1}^{n}(X_i - \overline{X})(Y_i - \overline{Y}) \tag{4.3c}$$

The regression line given by Eq. (4.1) (called Y on X regression) assumes that the independent variable X is known without error. This line will be different (steeper) for X on Y regression where Y is assumed to be known without error (Fig. 4.2). The differences in slope arise from the fact that Y on X minimizes the squared deviation parallel to the y-axis and vice versa. In practical applications, in particular when the functional relationship between the variables is assumed to be known based on physical understanding, the two regression lines can be viewed as limits when all the errors are attributed to one variable or the other. The reduced major axis (RMA) line lies between the two lines and assumes that the ratio of the error

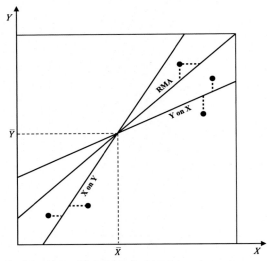

FIG. 4.2   An illustration of the different regression lines—Y on X, X on Y, and the RMA. *Modified from Doveton, J. H., 1994. Geologic Log Analysis Using Computer Methods. American Association of Petroleum Geologists, Tulsa, OK, p. 169.*

variances of the two variables is given by the ratio of their individual variances. The RMA line is obtained by minimizing the area between the points and the best-fit line, and the slope of the line is given by the ratio of the standard deviations of the two variables. All three lines pass through the point, $(\overline{X}, \overline{Y})$. When perfect correlation exists, all three lines coincide.

In general, the selection of the best-fit line depends on the specific application. If the goal is to simply predict one variable based on available measurements of the other, then a simple regression line may be adequate with the variable being predicted as the dependent variable. However, if the goal is to identify function or structural dependency between the variables, then the best-fit line should incorporate noise in both the variables, and the RMA line may be the preferred choice (Doveton, 1994).

4.2.2 Evaluating the Linear Regression Model

A common approach to examine the adequacy of the regression model is to determine how much of the variability in the dependent variable Y is explained by the regression line. To answer this, we start out by defining the following quantities (refer to Fig. 4.1):

$$e_i = Y_i - \hat{Y}_i = \text{residual describing the deviation of observed data from model predictions}$$

$$SS_E = \sum_{i=1}^{n} \left(Y_i - \hat{Y}_i\right)^2 = \sum_{i=1}^{n} e_i^2 = \text{residual sum of squares}$$

$$SS_R = \sum_{i=1}^{n} \left(\hat{Y}_i - \overline{Y}_i\right)^2 = \text{sum of squares due to regression}$$

$$S_{YY} = \sum_{i=1}^{n} \left(Y_i - \overline{Y}_i\right)^2 = \text{total sum of squares or sum of squares about the mean}$$

$$(4.4)$$

It can be shown that (see Haan (1986) for derivations)

$$S_{YY} = SS_E + SS_R \tag{4.5}$$

Thus, the total sum of squares has two components: the residual sum of squares and the sum of squares explained by the regression model. Clearly, if $SS_R \gg SS_E$, then the regression line can explain most of the variations in the dependent variable. A related measure is the coefficient of determination, R^2, which is defined as follows:

$$R^2 = \frac{SS_R}{S_{YY}} = 1 - \frac{SS_E}{S_{YY}} \tag{4.6}$$

The R^2 explains the fraction of the total sum of squares that is explained by the regression model. The range of R^2 will be between zero and one, with one being a perfectly fitting model ($SS_E = 0$) explaining all the variations in the dependent variable. It is important at this point to distinguish between this coefficient of determination and the coefficient of correlation, ρ, which are often used interchangeably. If X and Y are both random variables, then R^2 is the same as the square of the correlation coefficient between X and Y. In fact, the coefficient of correlation is an estimate of the population parameter for a joint normal distribution between X and Y, whereas R^2 makes no such assumption about the underlying distribution of the variables (Jensen et al., 1997).

The standard error of regression, also known as the standard error of the estimate, is given by the following:

$$\hat{\sigma} = \sqrt{\frac{SS_E}{(n-2)}} \tag{4.7}$$

Note that the square of the standard error of regression $\hat{\sigma}^2$ is an unbiased estimate of the error variance in $\varepsilon_i(0, \sigma^2)$. These are the variations in Y that are not explained by the regression model and can be described as the "true" noise in the data. In Eq. (4.7), the residual sum of squares is divided by $(n-2)$ to account for the fact that two degrees of freedom for error have been used up in estimating the slope and the intercept parameters in the regression model. One important underlying assumption in the regression model is that the unexplained variations in Y are independent (i.e., uncorrelated) and have a constant variance (i.e., homoscedastic). This is discussed further in the next section.

It is worth pointing out that the coefficient of determination in Eq. (4.6) is not an unbiased estimate. It needs to be adjusted for the loss of degrees of freedom (one for the mean in S_{YY} and two for the slope and intercept parameters in SS_E) in the same manner as in the calculation of standard error of regression in Eq. (4.7). This leads to the adjusted R^2 that is given by the following:

$$\text{Adjusted } R^2 = 1 - \frac{(n-1)}{(n-2)}(1 - R^2) \tag{4.8}$$

Note that the adjusted R^2 can be negative if R^2 is zero, and thus, X has no predictive value. Under such conditions, the mean model is clearly a better choice than the regression model.

4.2.3 Properties of the Regression Parameters and Confidence Limits

In order to develop a formalism for confidence intervals for the regression model and forecast, we will assume that ε_i, the deviation between the observed data and the unknown "true" model, follows a normal distribution with mean zero and variance σ^2, that is, $\varepsilon \sim N(0, \sigma^2)$. We can verify this assumption using diagnostic plots for the residuals, e_i, and ensuring that $e \sim N(0, \sigma^2)$. Such diagnostics plots are extremely useful for the analysis of regression models and checking its validity. We will illustrate this in the examples that will follow.

We can now make estimates of the standard error associated with the regression parameters, a and b (see Haan (1986) for derivations):

$$\hat{\sigma}_a = \hat{\sigma}\sqrt{\frac{1}{n} + \frac{\overline{X}^2}{S_{XX}}} \tag{4.9a}$$

$$\hat{\sigma}_b = \frac{\hat{\sigma}}{\sqrt{S_{XX}}} \tag{4.9b}$$

where S_{XX} is defined in Eq. (4.3c). Notice that the standard error of the parameters is directly proportional to the standard error of regression $\hat{\sigma}$. In other words, the noise in the data (as estimated by $\hat{\sigma}^2$) will equally impact all the regression parameters. It is important to note that as we obtain more data, the estimate of the standard error of regression $\hat{\sigma}$ will become more accurate, but there is no guarantee that the error itself will decrease. However, increasing the number of data points will reduce the standard error of the regression coefficient as can be seen from Eq. (4.9a).

Also, recall that the quantity S_{XX} in the denominator of Eqs. (4.9a), (4.9b) is a measure of the spread in X. Thus, everything else being equal, an experiment conducted with a wider range of X will result in less uncertainties in the regression coefficients and a more precise regression model.

If the regression model is correct, then the quantities $\hat{a}/\hat{\sigma}_a$ and $\hat{b}/\hat{\sigma}_b$ will be distributed as t distribution with $(n-2)$ degrees of freedom (Navidi, 2008). This allows to place confidence intervals in the regression parameters and examine the significance of the regression equation. The lower and upper confidence limits for the regression coefficients can be computed with appropriate t-values with $(n-2)$ degrees of freedom and the desired level of confidence α as follows:

$$L = \hat{a} - \hat{\sigma}_a t_{(1-\alpha/2),\,(n-2)}$$
$$U = \hat{a} + \hat{\sigma}_a t_{(1-\alpha/2),\,(n-2)} \tag{4.10a}$$

and

$$L = \hat{b} - \hat{\sigma}_b t_{(1-\alpha/2),\,(n-2)}$$
$$U = \hat{b} + \hat{\sigma}_b t_{(1-\alpha/2),\,(n-2)} \tag{4.10b}$$

It is worth pointing out here that the t distribution is very similar to the standard normal distribution but with a larger variance and a heavier tail (see Chapter 3). The larger variance comes from the fact that we replace the population variance with the sample variance in computing the t-values (in the same manner as z-values in the standard normal distribution), thus incorporating additional variability from samples.

In fact, as the number of degrees of freedom increases, the t distribution will approach the standard normal distribution.

4.2.4 Estimating Confidence Intervals for the Mean Response and Forecast

The confidence intervals on the regression line can be determined by calculating the standard error of the mean response, $\hat{Y} = \hat{a} + \hat{b}X$, for a given X. It is the error associated with the height of the regression line at a given location (Fig. 4.1). The standard error of the mean response can be estimated as follows (Haan, 1986):

$$\hat{\sigma}_{\overline{Y}} = \hat{\sigma}\sqrt{\frac{1}{n} + \frac{(X - \overline{X})^2}{S_{XX}}} \tag{4.11}$$

Notice that $\hat{\sigma}_{\overline{Y}}$ is minimum at $X = \overline{X}$, and it increases as we deviate from the mean. The confidence interval for the mean response can now be estimated using the standard error in the Eq. (4.11) with appropriate t-values with $(n-2)$ degrees of freedom and the desired level of confidence α as follows:

$$L_{\overline{Y}} = \hat{a} + \hat{b}X - \hat{\sigma}_{\overline{Y}}t_{(1-\alpha/2),\,(n-2)}$$
$$U_{\overline{Y}} = \hat{a} + \hat{b}X + \hat{\sigma}_{\overline{Y}}t_{(1-\alpha/2),\,(n-2)} \tag{4.12}$$

The confidence interval of individual predicted value of Y, also called the standard error of the forecast, will include both the unpredictable variability in Y given by the standard error of regression $\hat{\sigma}$ in Eq. (4.7) and the error in estimating the mean:

$$\hat{\sigma}_F = \sqrt{\hat{\sigma}^2 + \hat{\sigma}_{\overline{Y}}^2} = \hat{\sigma}\sqrt{1 + \frac{1}{n} + \frac{(X - \overline{X})^2}{S_{XX}}} \tag{4.13}$$

The lower and upper confidence limits for the forecast for Y for a given X can now be estimated using the standard error of forecast in Eq. (4.13) with appropriate t-values and confidence level as before:

$$L_F = Y - \hat{\sigma}_F t_{(1-\alpha/2),\,(n-2)}$$
$$U_F = Y + \hat{\sigma}_F t_{(1-\alpha/2),\,(n-2)} \tag{4.14}$$

Notice that both the confidence intervals for the mean response and the forecast are given for a specific value of X. By computing the confidence intervals for several values of X and connecting the points with a smooth curve give us the confidence bands and prediction bands for the regression model. For typical regression problems $(n > 30)$, the confidence intervals will generally equal to the forecast plus or minus two standard error of forecast for a 95% confidence level.

4.2.5 An Illustrative Example of Linear Regression Modeling and Analysis

Consider the 31-sample dataset (LINREG_FIG4.3.DAT) from an oil reservoir relating initial well potential that is the producing ability of the well (response variable) and net pay that is the net thickness of the productive interval (predictor variable). As shown in Fig. 4.3, the linear model is given by the following equation:

$$\hat{Y}_i = \hat{a} + \hat{b}X_i$$
$$\hat{a} = 2.063; \quad \hat{b} = 97.937 \tag{4.15}$$

How good is the model? Although Fig. 4.3 gives us a visual idea about the extent to which the model fits the data and captures the broad trend, a more informative diagnostic is to visualize the structure of the residuals. Recall that the linear model assumes that (i) the residuals are random and independent, (ii) the residuals have zero mean and the same (constant) variance, and (iii) the residuals are normally distributed.

Fig. 4.4 shows the diagnostic plots for the residuals for this example. In Fig. 4.4A, we expect the residuals to be symmetrical around zero and evenly spread for all values of the predictor variable (because of constant variance or homoscedasticity), which is indeed the case. Fig. 4.4B is a normal quantile plot of the residuals. A linear trend indicates that the residuals are, indeed, normally distributed.

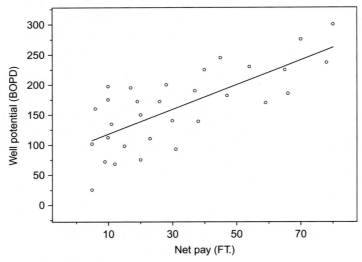

FIG. 4.3   Data and the fitted linear model.

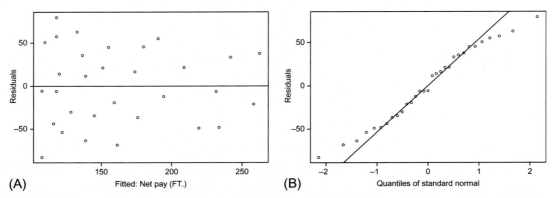

FIG. 4.4   Diagnostic plots for residuals, (A) residual versus the fitted values and (B) normal quantiles of the residuals.

Another useful diagnostic is a cross plot of the observed and the predicted response variable (Fig. 4.5). This plot is useful for detecting any systematic bias in the model, such as consistent over prediction or under prediction for a data range. We expect the plot to display uniform spread around the unit slope line to ensure that the unexplained variability in the data are indeed random and do not have an underlying structure, as is generally the case here.

We can construct confidence intervals for each point on the fitted regression line using Eq. (4.14) and smoothly connect them to generate the confidence band shown in Fig. 4.6.

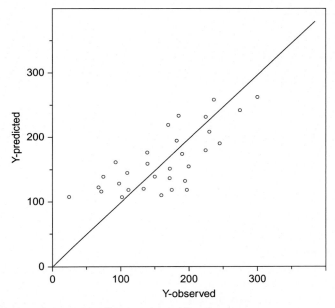

FIG. 4.5   Cross validation plot—observed versus the predicted response.

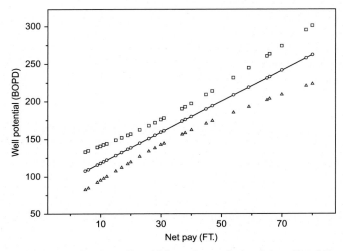

FIG. 4.6   Confidence bands given by the 95% confidence intervals for the predicted *Y*-values.

TABLE 4.1   **Summary of Simple Linear Regression Analysis**

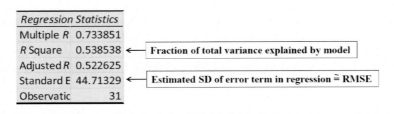

Regression Statistics	
Multiple R	0.733851
R Square	0.538538 ← Fraction of total variance explained by model
Adjusted R	0.522625
Standard E	44.71329 ← Estimated SD of error term in regression ≅ RMSE
Observatic	31

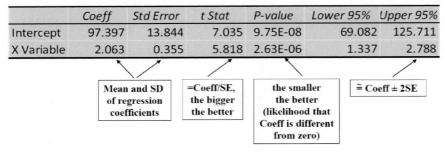

	Coeff	Std Error	t Stat	P-value	Lower 95%	Upper 95%
Intercept	97.397	13.844	7.035	9.75E-08	69.082	125.711
X Variable	2.063	0.355	5.818	2.63E-06	1.337	2.788

Mean and SD of regression coefficients

=Coeff/SE, the bigger the better

the smaller the better (likelihood that Coeff is different from zero)

≅ Coeff ± 2SE

As expected from Eq. (4.13), the uncertainty in forecast is smallest at the mean and increases as we deviate from the mean.

Finally, Table 4.1 provides a summary of the simple linear regression analysis. As explained before, the R^2 and the adjusted R^2 give the fraction of the total variance in Y explained by the linear regression model. The standard error of regression is an unbiased estimate of the variations in the data not explained by the model. The standard errors of the regression coefficients provide a measure of uncertainty in the estimated regression coefficients. The associated t-statistics and P-values are used for evaluating whether there is enough statistical evidence in the sample data for the regression model to be valid in general. The P-value for the slope parameter is particularly important in analyzing the regression model. If the P-value is sufficiently small for this parameter, we can infer that the likelihood of the slope being zero is very small, and the linear model is a reasonable choice.

4.3 MULTIPLE REGRESSION

4.3.1 Formulating and Solving the Multiple Regression Model

Multiple regression refers to situations when several independent variables are related to a single-dependent variable. Assume that the variable Y is related to p-independent variables, and we have n measurements. The simplest multiple linear regression model will be given by

$$Y_i = \beta_0 + \sum_{j=1}^{p} \beta_j X_{ij} + \varepsilon_i \quad i = 1.............n \tag{4.16}$$

The multiple regression model can also include powers of the independent variables and product terms describing variable interactions:

$$Y_i = \beta_0 + \beta_1 X_{1i} + \beta_2 X_{2i} + \beta_3 X^2_{1i} + \beta_4 X^2_{1i} + \beta_5 X_{1i} X_{2i} + \varepsilon_i \quad i = 1...........n \tag{4.17}$$

Eq. (4.17) is also called a linear model even though it contains nonlinear terms. The linearity here is with respect to the regression coefficients, β_j.

Just as in simple linear regression, the coefficients of multiple linear regression model can be obtained by minimizing the residual sum of squares. That is,

$$\text{Minimize} \sum_{i=1}^{n} e_i^2 \tag{4.18a}$$

The residual e_i is given by

$$e_i = Y_i - \left(\hat{\beta}_0 + \sum_{j=1}^{p} \hat{\beta}_j X_{ij} \right) \quad i = 1...........n \tag{4.18b}$$

The minimization is carried out in the usual manner, taking the partial derivatives of Eq. (4.18a) with respect to $\left(\hat{\beta}_0, \hat{\beta}_1 ....\hat{\beta}_p \right)$ and setting them to zero. This leads to (p + 1) normal equations that are solved for the (p + 1) regression coefficients. For a detailed derivation, we refer the reader elsewhere (Haan, 1986). Obviously, the results get more complicated compared with the simple linear regression and can be conveniently expressed in a matrix-vector form as follows:

$$\hat{\beta} = \left(H^T H \right)^{-1} H^T Y$$
$\hat{\beta}$ = vector of regression coefficients
Y = vector of measured dependent variable $\tag{4.19}$
H = matrix containing measured independent variables

Note that the specific structure of the H matrix will depend on the form of the multiple regression model.

The multiple regression model is a very powerful tool for data analysis, and a very broad range of problems can be handled using this technique. Quite often, nonlinear regression problem involving nonlinearity in the dependent variable can be reduced to multiple linear regression problem through appropriate transformation of the variables. These transformations can be parametric or nonparametric transformation. In particular, nonparametric transformation methods for multiple regression provide a flexible data-driven approach to interpreting complex relationships between variables in the absence of sound underlying physical models (Hastie and Tibshirani, 1990). The nonparametric regression methods will be discussed later in this chapter.

4.3.2 Evaluating the Multiple Regression Model

The evaluation of the multiple regression model is carried out very much like the simple linear regression before and uses the same sum of square quantities that are repeated below for completeness:

$e_i = Y_i - \hat{Y}_i$ = residual describing the deviation of observed data from model predictions

$$SS_E = \sum_{i=1}^{n} \left(Y_i - \hat{Y}_i \right)^2 = \sum_{i=1}^{n} e_i^2 = \text{residual sum of squares}$$

$$SS_R = \sum_{i=1}^{n} \left(\hat{Y}_i - \overline{Y}_i \right)^2 = \text{sum of squares due to regression}$$

$$S_{YY} = \sum_{i=1}^{n} \left(Y_i - \overline{Y}_i \right)^2 = \text{total sum of squares or sum of squares about the mean}$$

$$(4.20a)$$

Also, as in simple linear regression, the following identity holds (Navidi, 2008):

$$S_{YY} = SS_E + SS_R \qquad (4.20b)$$

Eq. (4.20b) is called the analysis of variance identity. These results are summarized in the form of an analysis of variance table for multiple regression as discussed later.

We can now define R^2 or the coefficient of determination describing the goodness-of-fit statistic in multiple regression:

$$R^2 = \frac{SS_R}{S_{YY}} = 1 - \frac{SS_E}{S_{YY}} \qquad (4.21)$$

The R^2 describes the proportion of total variance explained by the multiple regression model in the same way as in simple linear regression.

The standard error of regression is given by the following equation:

$$\hat{\sigma} = \sqrt{\frac{SS_E}{n-p-1}} \qquad (4.22)$$

Again, the expression for standard error of multiple regression is similar to that of standard linear regression. There are now $(n-p-1)$ degrees of freedom, because we are estimating $(p+1)$ regression coefficients rather than just 2.

In simple linear regression, we have seen that the t-statistic and P-values can be used to accept and to evaluate whether the slope parameter, β_1, is statistically indistinguishable from zero and, thus, examine the validity of the linear model. An analogous statistic for multiple regression is the F-statistic that is given by the following:

$$F = \frac{SS_R/p}{SS_E/(n-p-1)} \qquad (4.23)$$

The F-statistic can be used to test the hypothesis, $\beta_1 = \beta_2 = \ldots\ldots\ldots = \beta_n = 0$. In practice, we must reject this hypothesis based on the evidence from the data (as given by large observed values of the F-statistic or by small P-values) if the multiple regression model is appropriate.

The validity of the multiple regression can also be verified using diagnostic plots for the residuals versus fitted values as we have seen for simple linear regression. We expect the residuals to be independent random variables with zero mean and normally distributed. It is also recommended that we plot the residuals versus each of the independent variables to rule out any systematic trend.

4.3.3 How Many Terms in the Regression Model?

Deciding on the number of independent variables in a multiple regression model is often referred to as the model selection problem. The model selection process is guided by the principle of parsimony, which states that the model should contain the smallest number of variables required to fit the data. This requires balancing between model complexity (degrees of freedom) and goodness of fit.

The most widely used model selection process is a stepwise regression that involves evaluating the independent variables one at a time. The model selection rewards better fit but penalizes too many parameters based on the Akaike information criteria (AIC) (Navidi, 2008):

$$AIC = n \log (SS_E/n) + 2p$$
where
$n =$ number of observations
$p =$ number of model parameters
$SS_E =$ residual sum of squares

$$(4.24)$$

The goal here is to select the combination of independent variables that result in the minimum value of AIC. There are also other related measures of parsimony, for example, the Bayesian information criteria (BIC) given by Navidi (2008):

$$BIC = n \log (SS_E/n) + p \log (n)$$

$$(4.25)$$

An example of variable selection using the stepwise regression is discussed later in the chapter (see Section 4.5).

4.3.4 Analysis of Variance (ANOVA) Table

The analysis of variance table (shown in Table 4.2) is a commonly used summary of multiple regression results. It displays the partitioning of the sum of squares and the associated degrees of freedom. It is analogous to the summary of simple linear regression shown in Table 4.1. The degrees of freedom for regression are equal to the number of independent variables. The degrees of freedom for the residual error will be the number of observations minus the number of estimated parameters (the coefficients of the independent variables plus the intercept parameter). Thus, the total degrees of freedom will be the number of observations minus one. The mean squares are the sums of squares divided by their respective degrees of freedom and are used to calculate the F-statistic that is used to test the hypothesis that all the coefficients of the independent variables could be zero. A large value of F and small values of P will reject this hypothesis and establish the validity of the linear model.

TABLE 4.2   **Analysis of Variance Table**

Source	Degrees of Freedom	Sum of Squares	Mean Squares	F-Statistic	P-Value
Regression	p	SS_R	$MS_R = SS_R/p$	$F_{p,(n-p-1)} = MS_R/MS_E$	$0 \leq P \leq 1$
Residual error	$n-p-1$	SS_E	$MS_E = SS_E/(n-p-1)$		
Total	$n-1$				

4.3.5 An Illustrative Example of Multiple Regression Modeling and Analysis

The dataset used for this example is shown in Fig. 4.7A (MULTREG_FIG4-7.DAT). The dependent variable is the logarithm of permeability (PERM), whereas the independent variables are well-log responses, namely, gamma ray (GR) and bulk density (RHOB). The gamma-ray log is indicative of formation lithology, for example, sand versus shale, and the bulk density log is indicative of formation porosity. Thus, permeability is likely to be correlated to these well logs. The result of the multiple regression is shown in Fig. 4.7. The regression model is given by the following equation:

$$\ln(\text{Perm}) = -0.0215(\text{GR}) - 13.39(\text{RHOB}) + 35.175$$

Fig. 4.8 shows the diagnostic plots for the residuals versus the fitted model. As for simple linear regression, we expect the residuals to be symmetrical around zero and evenly spread for all values of the predictor variable (homoscedasticity). Fig. 4.8B is a normal quantile plot of the residuals. Again, the linear trend indicates that the residuals are, indeed, normally distributed. These results seem to indicate no serious violations of the regression model assumptions.

Another useful diagnostic plot for multiple regression is residual versus the independent variables as shown in Fig. 4.9 to detect any unexplained structure in the residual. For a good model, the residuals should be uncorrelated random noise with no systematic trend.

Finally, Table 4.3 provides a summary of the multiple linear regression analysis. As for simple linear regression, the R^2 and the adjusted R^2 give the fraction of the total variance in Y explained by the multiple linear model. The standard error of regression is an unbiased estimate of the variations in the data not explained by the model.

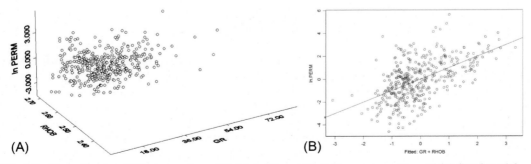

FIG. 4.7   **Example of multiple regression—(A) dataset for multiple linear regression and (B) fitted model.**

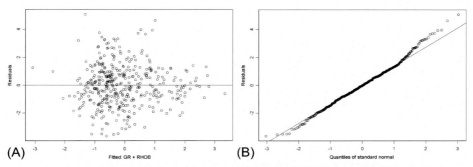

FIG. 4.8 Diagnostic plots for multiple regression—(A) residual versus fitted values and (B) normal quantile plot of the residuals.

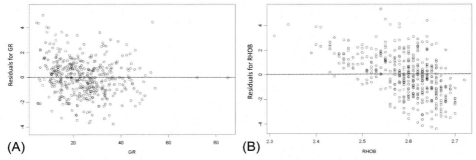

FIG. 4.9 Diagnostic plots for multiple regression—residuals versus the independent variables.

TABLE 4.3 **Summary of Multiple Linear Regression Analysis**

Regression Statistics

R square	0.309191667
Adjusted R square	0.305737625
Standard error	1.483332723
Observations	403

ANOVA

	df	SS	MS	F	P-Value
Regression	2	393.9193904	196.9597	89.51591	7.43898E-33
Residual	400	880.1103867	2.200276		
Total	402	1274.029777			

	Coefficients	Standard Error	t-Stat	P-Value
Intercept	35.17517182	2.862046273	12.29022	1.13E-29
X variable 1	−0.02151192	0.00666908	−3.22562	.00136
X variable 2	−13.3902369	1.115334064	−12.0056	1.44E-28

The analysis of variance (ANOVA) table displays the partitioning of the sum of squares and the associated degrees of freedom. The relatively large F-value and the very small P-value suggest that all of the independent variables are linearly related to the dependent variable in a statistically significant manner. The standard errors of the regression coefficients provide a measure of uncertainty in the estimated regression coefficients. Since the associated t-statistic is large and P-values are sufficiently small for all the parameters, we can infer that the coefficients of the independent variables are nonzero and the multiple linear regression model is a reasonable choice in this case.

4.4 NONPARAMETRIC TRANSFORMATION AND REGRESSION

4.4.1 Conditional Expectation and Scatterplot Smoothers

In general, the regression problem involves a set of predictors, for example, a p-dimensional random vector X and a random variable Y, which is called the response variable. The aim of regression analysis is to estimate the mean response or the *conditional expectation*, $E(Y \mid X_1, X_2, \ldots, X_p)$. We have seen that the multiple regression method requires a functional form to be presumed *a priori* for the regression surface, thus reducing the problem to that of estimating a set of parameters. Such parametric approach can be successful provided the model assumed is appropriate. When the relationship between the response and predictor variables is unknown or inexact, as is frequently the case for petroleum geoscience applications, parametric regression can yield erroneous and even misleading results. This is the primary motivation behind nonparametric regression techniques that make only few general assumptions about the regression surface (Friedman and Stuetzle, 1981).

The nonparametric transformation techniques generate regression relations in a flexible data-defined manner through the use of *scatterplot smoothers* and in doing so let the data suggest the functionalities. The most extensively studied nonparametric regression techniques are based on some sort of locally weighted averaging which takes the following form (Friedman and Silverman, 1989):

$$E(Y|X) \approx \sum_{i=1}^{N} H(X, X_i) Y_i \tag{4.26}$$

where $H(X, X')$, the local averaging or kernel function, usually has its maximum at $X = X'$ with its absolute value decreasing as $|X' - X|$ increases. A critical parameter in local averaging is the span or bandwidth $s(X)$ that is the size of the interval, centered at X', over which most of the averaging takes place as shown in Fig. 4.10.

Some examples of averaging function are given in Table 4.4.

In practice, the choice of the span or bandwidth is much more critical than the choice of the averaging function itself (Hastie and Tibshirani, 1990). A large bandwidth will generate a "smoother" curve (i.e., more bias), whereas a small bandwidth will introduce more variability (i.e., more variance). Thus, bias increases, whereas variance decreases with increasing bandwidth. This is illustrated in Fig. 4.11. The optimal choice of span is given by a compromise between bias and variance (Hastie and Tibshirani, 1990).

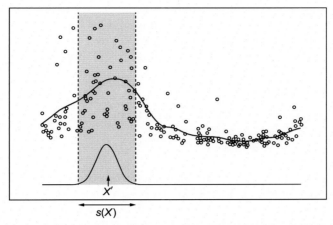

FIG. 4.10 **An illustration of scatterplot smoothing using a Gaussian kernel function.**

TABLE 4.4 **Some Examples of Local Averaging Functions**

Rectangular	$K(x) = 1$	$|x| < 1$
Triangular	$K(x) = 1 - |x|$	$|x| < 1$
Epanechnikov	$K(x) = 1 - x^2$	$|x| < 1$
Bisquare	$K(x) = (1 - x^2)^2$	$|x| < 1$
Tricube	$K(x) = (1 - x^3)^3$	$|x| < 1$
Triweight	$K(x) = (1 - x^2)^3$	$|x| < 1$
Gaussian	$K(x) = \exp(-(2.5x)^2 / 2)$	

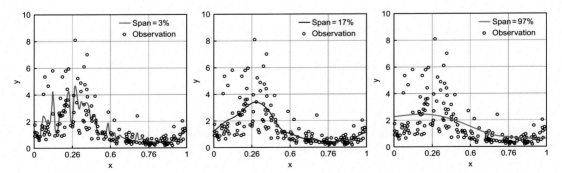

FIG. 4.11 **An illustration of the effects of span selection on scatterplot smoothing. Span size of 17% of the X-range is the optimum span here based on bias-variance tradeoff.**

Nonparametric regression methods can be broadly classified into those that do not transform the response variable (generalized additive models) and those that do (ACE and its variations). A brief discussion of these techniques follows. For further details, the reader is referred to Hastie and Tibshirani (1990), Buja et al. (1989), and Xue et al. (1997).

4.4.2 Generalized Additive Models

An additive regression model has the general form:

$$E(Y|X_1, X_2, \ldots, X_p) = \alpha + \sum_{l=1}^{p} \phi_l(X_l) + \varepsilon \qquad (4.27)$$

where X_l are the predictors and ϕ_l are functions of predictors. This can be viewed as an extension of the linear model in Eq. (4.16). Thus, additive models replace the problem of estimating a function of a p-dimensional variable X by one of estimating p separate one-dimensional functions, ϕ_l. Such models are attractive, if they can fit the data adequately, since they are often easier to interpret than a p-dimensional multivariate surface.

The technique for estimating ϕ_ls is called the *local scoring* algorithm and uses scatterplot smoothers, for example, a running mean, running median, running least-squares line, kernel estimates, or spline (see Buja et al. (1989) for a discussion of smoothing techniques). In order to explain the algorithm, let us consider the following simple model:

$$E(Y|X_1, X_2) = \phi_1(X_1) + \phi_2(X_2) \qquad (4.28)$$

Given an initial estimate $\phi_1(X_1)$, one way to estimate $\phi_2(X_2)$ is to smooth the residual $R_2 = Y - \phi_1(X_1)$ on X_2. With this estimate of $\phi_2(X_2)$, we can get an improved estimate $\phi_1(X_1)$ by smoothing $R_1 = Y - \phi_2(X_2)$ on X_1. The resulting iterative smoothing procedure is called *backfitting* (Hastie and Tibshirani, 1990) and forms the core of additive models.

In general, an algorithm for fitting a generalized additive model (GAM) consists of a hierarchy of three modules: (i) the *scatterplot smoothers* that can be thought of as a general regression tool for fitting functional relationship between response and predictor variables, (ii) a *backfitting* algorithm that cycles through the individual terms in the additive model and iteratively updates each by smoothing suitably defined partial residuals, and (iii) a *local scoring* algorithm that utilizes an iteratively reweighted least-squares procedure to generate a new additive predictor. A step-by-step procedure for the GAM can be found in Hastie and Tibshirani (1990).

4.4.3 Response Transformation Models: ACE Algorithm and Its Variations

The response transformation models generalize the additive model by allowing for a transformation of the response variable Y. The models have the following general form:

$$\theta(Y) = \sum_{l=1}^{p} \phi_l(X_l) + \varepsilon \qquad (4.29)$$

The main motivation behind response transformation is that often, a simple additive model may not be appropriate for $E(Y \mid X_1, X_2, \ldots, X_p)$, but may be quite appropriate for $E\{\theta(Y) \mid X_1, X_2, \ldots, X_p\}$. An example of such model is the ACE algorithm and its modifications.

The ACE algorithm, originally proposed by Breiman and Friedman (1985), provides a method for estimating optimal transformations for multiple regression that result in a maximum correlation between a dependent (response) random variable and multiple independent (predictor) random variables. Such optimal transformations can be derived by minimizing the variance of a linear relationship between the transformed response variable and the sum of transformed predictor variables as shown in Fig. 4.12.

For a given set of response variable Y and predictor variables $X_1,..., X_p$, the ACE algorithm starts out by defining arbitrary measurable mean-zero transformations $\theta(Y), \phi_1(X_1),...........$ $\phi_p(X_p)$. The error (e^2) not explained by a regression of the transformed dependent variable on the sum of transformed independent variables is under the constraint, $E\left[\theta^2(Y)\right] = 1$):

$$e^2\left(\theta, \phi_1, ..., \phi_p\right) = E\left\{\left[\theta(Y) - \sum_{l=1}^{p} \phi_l(X_l)\right]\right\}^2 \tag{4.30}$$

The minimization of e^2 with respect to $\phi_1(X_1), ...\phi_p(X_p)$ and $\theta(Y)$ is carried out through a series of single-function minimizations. Two basic mathematical operations involved in here are conditional expectations and iterative minimization and, hence, the name *alternating conditional expectations*. The final $\phi_l(X_l), l=1,...,p$ and $\theta(Y)$ after the minimization are estimates of optimal transformation, $\phi^*_l(X_l), l=1,...,p$ and $\theta^*(Y)$. In transformed space, the response and predictor variables will be related as follows:

$$\theta^*(Y) = \sum_{l=1}^{p} \phi_l^*(X_l) + \xi \tag{4.31}$$

The optimal transformations are derived solely based on the datasets and can be shown to result in a maximum correlation in the transformed space (Breiman and Friedman, 1985). The transformations do not require a priori assumptions of any functional form for the response or predictor variables and thus provide a powerful tool for exploratory data analysis and correlation. The transformation $\theta(Y)$ is assumed to be strictly monotone (and thus invertible), and the conditional expectations are approximated using scatterplot smoothers. A step-by-step procedure for the ACE model and its variations can be found in Breiman and Friedman (1985) and Hastie and Tibshirani (1990).

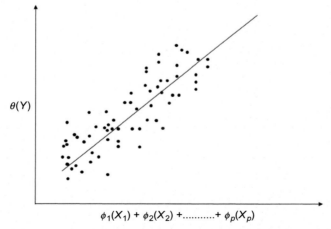

FIG. 4.12   **Optimal transformation for multiple regression.** *Based on Breiman, L., Friedman, J. H. 1985. Estimating optimal transformations for multiple regression and correlation. J. Am. Stat. Assoc. 80, 580.*

4.4.4 Data Correlation via Nonparametric Transformation

Nonparametric transformation techniques offer a flexible and data-driven approach to building correlation without a priori assumptions regarding functional relationship between response and predictor variables. The following equation is used to estimate or predict dependent variable, Y_i^{pre}, for any given data point $\{X_{1i}, X_{2i}........X_{pi}\}$ involving p- independent variables:

$$Y_i^{pre} = \theta^{*-1}\left[\sum_{l=1}^{p} \phi_l^*(X_{li})\right] \tag{4.32}$$

The calculation involves p forward transformations of $\{X_{1i}, X_{2i}........X_{pi}\}$ to $\{\phi_1^*(X_{1i}),...,\phi_p^*(X_{pi})\}$ and a backward transformation (Eq. 4.32). By restricting the transformation of the response variable to be monotone, we can ensure that θ^* is invertible.

The power of nonparametric transformations as a tool for correlation also lies in their ability to handle variables of mixed type. For example, we can easily incorporate categorical variables such as rock types and lithofacies into the correlation without additional complications (Datta-Gupta et al., 1999). Another important application of nonparametric transformation is function identification for data correlation using multiple regression as illustrated in the example below.

This example is designed after Breiman and Friedman (1985) and demonstrates the ability of nonparametric transformations to identify functional relationship during multiple regression. A dataset with 200 observations is simulated from the following model:

$$y_i = \exp\left[\sin(2\pi x_i) + \varepsilon_i/2\right] \quad (1 \le i \le 200) \tag{4.33}$$

where x_i is drawn from a uniform distribution $U(0,1)$ and ε_i is independently drawn from a standard normal distribution $N(0,1)$ (SCATTER_FIG4-13.DAT).

Fig. 4.13 is a scatterplot of y_i versus x_i. The plot itself does not reveal a functional relationship between the dependent and independent data observations. In this situation, the direct use of parametric regression is difficult and requires trial and error. If we take logarithm of both sides of Eq. (4.33), we obtain a linear relationship between $\ln(y_i)$ and $\sin(2\pi x_i)$ as follows:

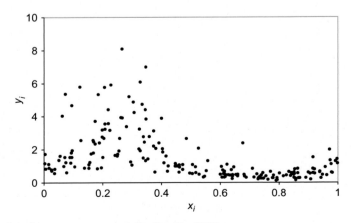

FIG. 4.13   Scatterplot of y_i versus x_i generated using Eq. (4.33).

$$\ln(y_i) = \sin(2\pi x_i) + \varepsilon_i/2 \tag{4.34}$$

Thus, the optimal transformations for linear regression will have the following forms:

$$\begin{aligned}
\theta^*(y_i) &= \ln(y_i) \\
\phi^*(x_i) &= \sin(2\pi x_i)
\end{aligned} \tag{4.35}$$

To demonstrate that the ACE algorithm can estimate the above optimal transformations, we applied the algorithm to the synthetic dataset in Fig. 4.13. Fig. 4.14A and B shows the optimal transformations of y_i and x_i derived by the ACE algorithm. Clearly, ACE is able to identify the logarithmic function as the optimal transformation of the dependent variable and the sine function as the optimal transformation of the independent variable. Fig. 4.15 shows a plot of $\theta^*(y_i)$ versus $\phi^*(x_i)$. A linear regression on the transformed data yields the following:

$$\theta^*(y_i) \approx 1.093 \, \phi^*(x_i) \tag{4.36}$$

which is a very close estimate of $\theta^*(y_i) = \phi^*(x_i)$ indicating that the transformations are, indeed, optimal.

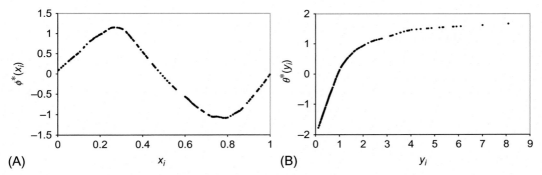

(A)

(B)

FIG. 4.14 **(A) Optimal transformation of x_i by ACE and (B) optimal transformation of y_i by ACE.**

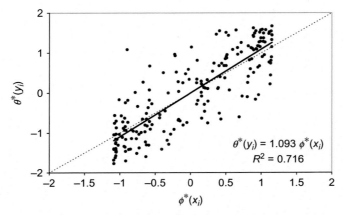

FIG. 4.15 **Optimal transformation of y_i versus optimal transformation of x_i by ACE. The solid straight line represents linear regression of the data.**

4.5 FIELD APPLICATION FOR NONPARAMETRIC REGRESSION: THE SALT CREEK DATA SET

4.5.1 Dataset Description

We demonstrate the application of nonparametric regression, specifically the ACE method and variable selection using a field example. Our goal here is to predict permeability using a suite of well logs in the Salt Creek Field Unit (SCFU), a highly heterogeneous carbonate reservoir in the Permian Basin, West Texas (Fig. 4.16). The data used in this analysis belong to seven wells with cores and measured permeabilities for the cored interval (Lee et al., 2002). A suite of seven well logs (GR, LLD, MSFL, DT, NPHI, RHOB, and PEF) is used to correlate core permeabilities with well-log response. Out of the seven cored wells, one well (G517) is left out to verify the correlations using blind tests. This application is well suited for nonparametric regression as the functional relationship between permeability and well logs is typically not known a priori. Nonparametric regression methods can be used to develop this functional relationship in a data-driven manner as illustrated in this example.

4.5.2 Variable Selection

We start with the full set of seven well-log data and select the appropriate combination of independent variables using a stepwise regression as shown in Table 4.5. Stepwise algorithm includes backward elimination and forward selection. Backward elimination is the simplest of all variable selection procedures and is illustrated here. The procedure proceeds as follows:

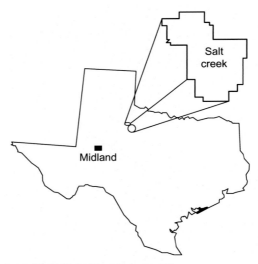

FIG. 4.16   **Location of Salt Creek Field Unit, Kent County, TX.**

TABLE 4.5   **Variable Selection Using Stepwise Algorithm**

Step	Add/Delete	GR	LLD	MSFL	DT	NPHI	RHOB	PEF	RSS	AIC
Step 1	Full	X	X	X	X	X	X	X	1073	383.7
Step 2	Initial	X	X	X	X	X	X	X	1073	383.7
	-GR		X	X	X	X	X	X	1153	420.5
	-LLD	X		X	X	X	X	X	1082	386.2
	-MSFL	X	X		X	X	X	X	1078	384.0
	-DT	X	X	X		X	X	X	1074	381.8
	-NPHI	X	X	X	X		X	X	1292	482.4
	-RHOB	X	X	X	X	X		X	1093	391.8
	-PEF	X	X	X	X	X	X		1114	402.1
Step 3	Initial	X	X	X		X	X	X	1074	381.8
	-GR		X	X		X	X	X	1151	417.5
	-LLD	X		X		X	X	X	1085	385.7
	-MSFL	X	X			X	X	X	1076	381.2
	-NPHI	X	X	X			X	X	1343	501.4
	-RHOB	X	X	X		X		X	1098	392.2
	-PEF	X	X	X		X	X		1114	400.0
Step 4	Initial	X	X			X	X	X	1076	381.2
	-GR		X			X	X	X	1156	418.1
	-LLD	X				X	X	X	1087	384.6
	-NPHI	X	X				X	X	1353	503.8
	-RHOB	X	X			X		X	1099	390.7
	-PEF	X	X			X	X		1119	400.4
Optimum		X	X			X	X	X	1076	381.2

(1) Fit the data with all the well logs and calculate the AIC; (2) delete variables one by one and recompute AIC criteria. If there is a smaller AIC value, select that model with least AIC value and repeat step (2). If no model has smaller AIC than the initial model, stop the stepwise procedure and select initial model as the optimal model.

Notice that out of the seven well logs we started with, two well logs (DT and MSFL) are removed by the stepwise regression. This is partly because other well logs (RHOB, NPHI, and LLD) already contain equivalent information.

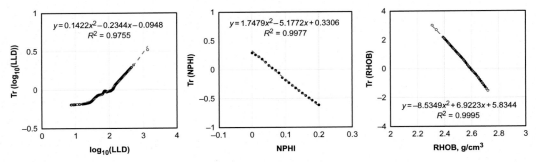

FIG. 4.17   Optimal transformations of some of independent variables as obtained by the ACE algorithm.

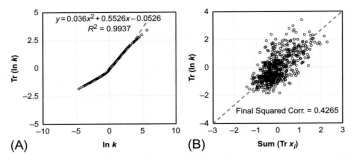

FIG. 4.18   (A) Optimal transformations of the dependent variable and (B) optimal correlation in the transformed space.

4.5.3 Optimal Transformations and Optimal Correlation

Fig. 4.17 shows the optimal transformations of some of the well logs obtained using the ACE algorithm. Also shown in Fig. 4.18 are the transformation of the dependent variable (log permeability) and the optimal correlation between the transformed dependent variables and the sum of transformed independent variables as given in Eq. (4.31).

Because these transformations are generally restricted to be smooth, we can fit them with simple polynomials to build a predictive model for permeability and well logs as shown below. For example, the transformation of the well logs is fitted with the following equation:

$$\phi^*(GR) = 0.0007GR^2 - 0.0605GR + 0.9493$$

$$\phi^*(\log LLD) = 0.1422(\log LLD)^2 - 0.2344\log(LLD) - 0.0948$$

$$\phi^*(NPHI) = 1.7479NPHI^2 - 5.1772NPHI + 0.3306$$

$$\phi^*(PEF) = -0.0058(PEF)^2 + 0.0355PEF + 0.0152$$

$$\phi^*(RHOB) = -3.5349(RHOB)^2 + 6.9223RHOB + 5.8344$$

The log permeability for a given set of well-log values can now be obtained by first computing their respective transformations using the equations above, summing the

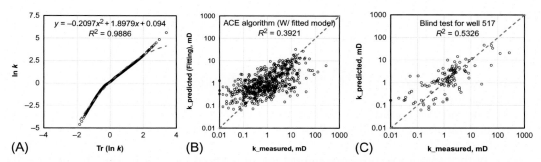

FIG. 4.19   Inverse transformation for predicting logarithm of permeability (A), measured versus the predicted data (B), and measured versus the predicted data for a blind well not used in the correlation (C).

transformations, followed by an inverse transformation as given in Eq. (4.32). The inverse transformation is shown in Fig. 4.19A and is given below:

$$\ln(k) = -0.2097\left[\sum\varphi^*(x_i)\right]^2 + 1.8979\sum\varphi^*(x_i) + 0.094 \tag{4.37}$$

The predicted permeabilities using Eq. (4.37) for the Salt Creek example are shown in Fig. 4.19B. In Fig. 4.19C, we have also shown the predicted permeabilities based on well logs versus the measured permeabilities for the blind well G517. The measured and predicted values appear to be evenly distributed around the unit slope line, indicating no systematic bias in the regression model.

The power of the nonparametric regression lies in the fact that it does not require any a priori assumption regarding the functional form between the dependent and independent variable. This is extremely useful because for earth sciences applications, such functional forms are often not known. We have seen that the ACE algorithms generate the transformations in a data-driven manner, and the transformations can be fitted with simple equations to develop predictive equations. The reader can reproduce the results in this field example using the software GRACE, and the Salt Creek field data (SALT-CREEK.DAT) made available in the online resources for this book.

4.6 SUMMARY

In this chapter, we have introduced data modeling and analysis using linear regression, multiple regression, and nonparametric regression that generates the regression relation in a data-driven manner without prior assumption regarding functional forms. Although we have presented the relevant equations necessary for modeling and interpretation of the results, our emphasis has been on the application and analysis rather than derivation of the equations. We have illustrated the power and utility of regression modeling using simple illustrative examples. Finally, a field application of nonparametric regression demonstrates the versatility of the method as a predictive tool.

Exercises

1. Show that the following relationship in Eq. (4.5) holds.

$$S_{YY} = SS_E + SS_R$$

(Hint: $\left(Y_i - \overline{Y}\right) = \left(Y_i - \hat{Y}_i\right) + \left(\hat{Y}_i - \overline{Y}\right)$. Square both sides, sum over all observations and manipulate)

2. Compute the 95% confidence interval on the permeability-thickness estimated from the pressure transient data with constant rate drawdown test. In this example, the pressure data fluctuate due to the difficulty to control the drawdown rate to be constant. Note that slope of the bottomhole pressure against log Δt is given by $-162.6\ qB\mu/kh$. Also, $q = 1000$ stb/day, $B = 1.0$ rbbl/stb, $\mu = 1$ cp. Calculate R^2 and adjusted-R^2 for linear regression.

Δt (h)	P_{wf} (psia)
0.000	5000.0
1.000	4841.9
2.000	4839.3
3.000	4826.0
4.000	4824.8
5.000	4835.3
6.000	4830.7
8.000	4825.2
12.000	4822.8
16.000	4813.0
20.000	4812.2
24.000	4793.7

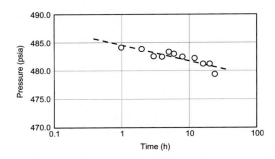

3. Using the following dataset, build a multivariate linear regression model to predict the bubble point pressure (dependent variable) using API, gas gravity, solution gas-oil-ratio (R_s), and reservoir temperature (T_{res}) as independent variables. Plot the residuals against the independent variables. Are there any points that are possible outliers? Do the residual plots have any patterns that suggest that the fitted regression model is not appropriate?

Oil Gravity (API)	Gas Gravity (SG)	T_{res} (°F)	R_s (scf/stb)	P_b (psi)
48.0	0.801	215	1512	3384
42.7	0.808	299	1138	3699
42.5	0.809	297	1472	4125
45.6	0.835	244	1534	3187
50.4	0.789	275	1194	3005
41.8	0.815	280	1567	4264
38.8	0.752	230	1492	4433
47.2	0.852	255	1966	3424
35.2	0.844	216	448	1704
26.2	0.705	192	1007	5297
34.4	0.617	215	957	4918
42.2	0.861	220	739	2421

4. Compute the R^2 and adjusted-R^2 values, standard error of regression and F-statistic for Exercise 3.

5. Compute the AIC for multivariate linear regression in Example 2. Furthermore, repeat the multivariate linear regression by removing one of the independent variables at a time as shown. Comment if using all the four independent variables is appropriate or not.

Oil Gravity	Gas Gravity	T_{res}	GOR	SSE (psi^2)	AIC
x	x	x	x	8.70E+05	66.3
x	x	x		4.00E+06	72.3
x	x		x	1.33E+06	66.5
x		x	x	3.65E+06	71.8
	x	x	x	2.50E+06	69.8

6. Using the dataset "SCATTER_FIG4-13.DAT," test three different smoothers (running mean, running median, and Gaussian) for three different span sizes as illustrated in Fig. 4.11.

7. Using the dataset "SALT-CREEK.DAT," perform nonparametric regression using the ACE algorithm.

   (1) Plot core permeability (ln kg) versus core porosity (POR) and do the linear regression. Do you think predicting the permeability only based on the porosity is a good idea for this particular case?
   (2) With variables selected in Table 4.5, perform the multivariate linear regression.
   (3) With variables selected in Table 4.5, do the Nonparametric regression using ACE algorithm and reproduce the results of Figs. 4.17–4.19.
   (4) Using the dataset for a blind test (SALT-CREEK-G517.DAT), predict the permeability with the three models: linear regression, multivariate linear regression, and nonparametric regression. See and compare R^2 values.

8. Consider the dataset relating Initial Well Potential to Net Pay (LINREG_FIG4-3.DAT).

(a) Fit a linear regression model with initial well potential as the response variable and the net pay as the predictor variable.

(b) What does the model predict about increase in initial well potential as the net pay increases by 10 ft?

(c) What are the estimates of initial well potential for net pay of 50 ft and 100 ft?

(d) What is the estimate of the error variance?

(e) What is the standard error of the slope parameter, $\hat{b}$?

(f) Construct a two-side 95% confidence interval for the slope parameter $\hat{b}$?

(g) Construct a two-side 95% confidence interval for the estimate of initial well potential for net pay of 50 ft and 100 ft.

References

Albertoni, A., Lake, L.W., 2003. Inferring Interwell Connectivity Only From Well-Rate Fluctuations in Waterfloods. Society of Petroleum Engineers. https://doi.org/10.2118/83381-PA.

Breiman, L., Friedman, J.H., 1985. Estimating optimal transformations for multiple regression and correlation. J. Am. Stat. Assoc. 80, 580.

Buja, A., Hastie, Trevor, Tibshirani, Robert, 1989. Linear smoothers and additive models. Ann. Stat. 17, 453–510.

Datta-Gupta, A., Xue, Guoping, Lee, Sangheon, 1999. Non-parametric transformations for data correlation and integration: from theory to practice. In: Schatzinger, R., Jordan, J. (Eds.), Recent Advances in Reservoir Characterization. American Association of Petroleum Geologists, Tulsa.

Doveton, J.H., 1994. Geologic Log Analysis Using Computer Methods. American Association of Petroleum Geologists, Tulsa, OK. p. 169.

Friedman, J.H., Silverman, B.W., 1989. Flexible parsimonious smoothing and additive modeling. Technometrics 31 (1), 3–20.

Friedman, J.H., Stuetzle, W., 1981. Project pursuit regression. J. Am. Stat. Assoc. 76 (376), 817–823.

Haan, C.T., 1986. Statistical Methods in Hydrology. Iowa University Press, Ames. 376.

Hastie, T., Tibshirani, R., 1990. Generalized Additive Models. Chapman and Hall, London. 335.

Jensen, J.L., Lake, L.W., Corbett, P.W.M., Goggin, D.J., 1997. Statistics for Petroleum Engineers and Geoscientists. Prentice Hall, New Jersey. 390.

LaFollette, R.F., Izadi, G., Zhong, M., 2014. Application of Multivariate Statistical Modeling and Geographic Information Systems Pattern-Recognition Analysis to Production Results in the Eagle Ford Formation of South Texas. Society of Petroleum Engineers. https://doi.org/10.2118/168628-MS.

Lee, S.H., Khraghoria, A., Datta-Gupta, A., 2002. Electrofacies characterization and permeability predictions in carbonate reservoirs: role of multivariate analysis and non-parametric regression. SPE Reserv. Eval. Eng. 5 (3).

Navidi, W., 2008. Statistics for Engineers and Scientists. McGraw-Hill, Boston. 901.

Voneiff, G., Sadeghi, S., Bastian, P., Wolters, B., Jochen, J., Chow, B., et al., 2013. A Well Performance Model Based on Multivariate Analysis of Completion and Production Data from Horizontal Wells in the Montney Formation in British Columbia. Society of Petroleum Engineers. https://doi.org/10.2118/167154-MS.

Wendt, W.A., Sakurai, S., Nelson, P.H., 1986. Permeability prediction from well logs using multiple regression. In: Lake, L.W., Carroll Jr., H.B., (Eds.), Reservoir Characterization. Academic Press, Inc., Orlando, FL, p. 659.

Xue, G., Datta-Gupta, A., Valko, P., Blasingame, T., 1997. Optimal transformations for multiple regression: application to permeability estimation from well logs. SPE Form. Eval. 12 (2), 85–94.

Multivariate Data Analysis

In this chapter, we introduce multivariate data analysis techniques, viz. principal component analysis, cluster analysis, and discriminant analysis in the context of data partitioning and pattern recognition for multiple regression. After introducing the concepts using a simple example, we discuss in detail the application of these techniques to the Salt Creek field data introduced in the previous chapter.

5.1 INTRODUCTION

In the previous chapter, we introduced multivariate regression techniques involving two or more variables. Before embarking on an analysis involving large number of variables, we might want to first examine if there are any underlying data structure or patterns that we can exploit to improve and sometimes simplify the analysis. A common approach will be to graphically visualize the data cloud that is limited to three variables. Often, a fourth dimension can be added by varying the type and size of symbols, but that is our limit for graphic visualization. For high-dimensional datasets, an alternative approach is to reduce the dimensionality of the data with minimum loss of important attributes, for example, data variance. Multivariate data analysis techniques allow us to accomplish these goals. Essentially, we define a smaller number of linear combination of the original data, called principal components that allow for data visualization and pattern recognition in a reduced dimensional space. The pattern recognition or classification techniques can be either "supervised" or "unsupervised." In the unsupervised classification techniques, commonly known as cluster analysis, we partition the data into relatively "homogeneous" entities based on the characteristics of the data, without resorting to prior information. In the supervised pattern-recognition method, also known as discriminant analysis, we assign group membership to a given dataset based on a prior classification. Multivariate data analysis by itself is a vast topic, and several excellent references are available on this topic (Hastie et al., 2008; Davis, 1986; Mardia et al., 1979). There are numerous applications of multivariate data analysis in petroleum engineering and geosciences. Some examples include formation evaluation (Hempkins, 1978), drilling (Hempkins et al., 1987), geophysical data analysis (Mwenifumbo, 1993), well completion optimization (Nitters et al., 1995), reservoir characterization (Scheevel and Payrazyan, 2001), and candidate selection for enhanced oil recovery (Siena et al., 2016).

5.2 PRINCIPAL COMPONENT ANALYSIS

The major motivation of principal component analysis (PCA) is to reduce the dimensionality of multivariate data involving large number of observations without significant loss of underlying information content. The principal components define a variance maximizing mutually orthogonal coordinate system and provide a convenient mechanism for visualizing and analyzing the data. Typically, the first few principal components are adequate to explain the majority of the variability in the data, and hence, PCA is used to represent the data in a reduced dimensional and mutually independent space. The principal component loadings relate the principal components to the original data. They provide a summary of the influence of the original variables on the principal components and constitute a useful basis for the interpretation of the data using the principal components. Principal components constitute an alternative form of displaying the data, thereby allowing better knowledge of its structure without changing the information.

5.2.1 Computing the Principal Components

The principal components can be thought of as mutually independent surrogate variables obtained by a coordinate transformation and projection of the data in a newly defined orthogonal coordinate system. The coordinate system is represented by the eigenvectors of the data covariance matrix, and the principal components are a weighted linear combination of the original data.

Suppose we have a dataset, $\mathbf{X}_{n \times p}$, in which the element x_{ij} corresponds to the data at the ith row and the jth column ($i = 1, \ldots, n$, and $j = 1, \ldots, p$). As we have seen in the previous chapters, while dealing with variances and covariances, it is more convenient to work with variables that represent the deviations from their respective means. Let $\mathbf{Z}_{n\times p}$ be the matrix of deviations from the mean for each column in $\mathbf{X}_{n\times p}$. The covariance matrix of the dataset $\mathbf{X}$ can then be defined by

$$\mathbf{\Sigma} = \mathbf{Z}^\mathsf{T}\mathbf{Z}/(n-1) \tag{5.1}$$

Note that the total variance S of the dataset will be given by the sum of the diagonal elements of the covariance matrix, that is, $S = \text{Trace}(\mathbf{\Sigma})$. Also, the covariance matrix is symmetrical and nonnegative definite, and its eigenvalues will be ≥ 0. Derivation of the principal components follows from the matrix property that a symmetrical and nonsingular matrix $\mathbf{\Sigma}$ can be factored into a combination of diagonal and orthogonal matrices via a spectral decomposition (Strang, 1998):

$$\mathbf{\Sigma} = \mathbf{Q}^\mathsf{T}\mathbf{\Lambda}\mathbf{Q} \tag{5.2}$$

where $\mathbf{\Lambda} = \text{diag}(\lambda_1, ..., \lambda_p)$, a diagonal matrix of eigenvalues of $\mathbf{\Sigma}$, and $\mathbf{Q}$ is a $p \times p$ orthogonal matrix whose column vectors consist of eigenvectors associated with the eigenvalues, $\lambda_1, ..., \lambda_p$, respectively.

Note that a row vector of the multidimensional data $\mathbf{X}$ can be represented as a point in p-dimensional space; hence, $\mathbf{X}$ forms a cloud of points in the multidimensional space. The eigenvectors are the principal axes of the cloud, and the matrix $\mathbf{Q}$ is used to transform the original data into principal components, $\mathbf{Y}_{n\times p}$:

$$\mathbf{Y} = \mathbf{ZQ} \tag{5.3}$$

If we represent Eq. (5.3) in terms of column vectors $\mathbf{Y} = [y_1, y_2 ... y_p]$, $\mathbf{Z} = [z_1, z_2 ... z_p]$, and $\mathbf{Q} = [q_1, q_2 ... q_p]$, then some important properties of the principal components can be summarized as follows:

   (i) y_i and y_j are independent (uncorrelated) for $i \neq j$.
   (ii) The magnitudes of eigenvalues of $\sum$ are given by $\lambda_1 \geq \lambda_2 \geq \cdots \geq \lambda_p$.
   (iii) The variance of the principal components is given by $\text{Var}(y_i) = \lambda_i$.

   (iv) Total variance of all principal components is given by $\sum_{i=1}^{p} \text{Var}(y_i) = \sum_{i=1}^{p} \lambda_i = \text{Trace}(\mathbf{\Sigma}) = S$.

From (iii) and (iv), we can see that the ratio $\lambda_i/\Sigma\lambda_i$ describes the proportion of the total data variance S explained by the principal component i.

The principal components are not scale-invariant and will be different depending on whether they are computed using the unscaled covariance matrix or the scaled correlation matrix. In general, one can use the covariance matrix when the original observations are on the same scale. However, the common practice is to use the correlation matrix by rescaling the variables by subtracting the mean and dividing by the standard deviation, particularly when the observations are of different types, for example, well-log measurements reflecting different subsurface properties. Fig. 5.1 illustrates the concept of principal components using a dataset consisting of two variables $x1$ and $x2$.

From Fig. 5.1, we can see that the principal component analysis involves a coordinate rotation and a projection of the data in the new coordinates. The weighting factors for the principal components are given by the eigenvectors of the correlation matrix of the original variables, and the relative variance of the principal components is given by the eigenvalues

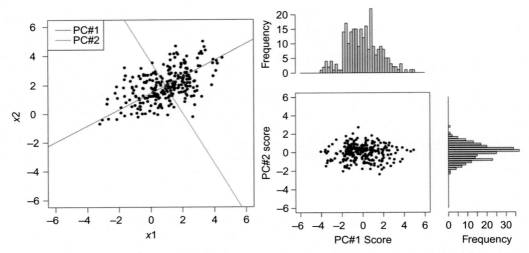

FIG. 5.1 **A simple illustration of the principal component analysis. It involves a coordinate rotation with the first principal component aligned in the direction of maximum variability in the data.**

of the correlation matrix. The weighting factors or coefficients of the principal component transformation are called "principal component loadings." The weighting factors describe the influence of the original variables on the principal components and thus can provide a useful basis for data interpretation. Because the eigenvalues typically decay rapidly, the first few principal components are often adequate to explain variability of the dataset. Hence, principal component analysis provides a mechanism of reducing the dimensionality of the original dataset without significant loss of data variability. There are several criteria available to decide on the number of principal components to be retained for data analysis. A common practice is to prespecify a certain percentage of variance to be preserved (e.g., >90%) and select enough principal components to satisfy the requirements. Another approach is to exclude the principal components associated with eigenvalues less than the average of all eigenvalues. We illustrate the steps in principal component analysis using a simple example below.

5.2.2 An Illustrative Example of the Principal Component Analysis

Consider the dataset displayed in Fig. 5.2 (MULTIVAR_FIG5-2.DAT). There are 29 data points here with three variables $X1$, $X2$, and $X3$.

The first step in principal component analysis will be to rescale the data. This involves normalizing each variable by subtracting its mean and dividing by its standard deviation. This step puts all the variables in an even footing by making them dimensionless with zero mean and unit variance. Next, we construct the correlation matrix. This is followed by a spectral decomposition of the correlation matrix to obtain the eigenvalues and the associated eigenvectors. These results are summarized in Table 5.1.

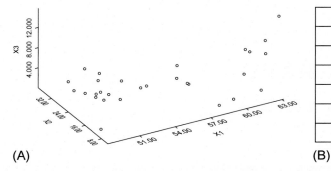

X1	X2	X3
59.31	7.44	1.472
57.63	5.21	2.027
60.25	5.59	10.879
61.69	5.98	9.562
63.19	7.86	15.802
....	....	....

(A)   (B)

FIG. 5.2   (A) A display of the cloud of the three-dimensional dataset and (B) a few sample data values in the cloud.

TABLE 5.1   Correlation Matrix and its Spectral Decompostion for the Example Data

	X1	X2	X3
Data correlation matrix			
X1	1.00	−0.731	0.726
X2	−0.731	1.00	−0.669
X3	0.726	−0.669	1.00

	Eigenvalues	Percentages	Cum. Percentage
Eigenvalues and the associated variances			
λ_1	2.42	80.5857	80.5857
λ_2	0.33	11.0426	91.6282
λ_3	0.25	8.3718	100

	EV-1	EV-2	EV-3
Eigenvectors defining the coefficients of principal components			
X1	0.5879	0.0269	0.8085
X2	−0.5727	−0.692	0.4395
X3	0.5713	−0.7214	−0.3915

A graphic representation of the results of principal component analysis is shown in Fig. 5.3. The screeplot displays the eigenvalues in Table 5.1 against their indexes. The plot also shows the proportion of the total variance explained by each principal component. A sharp drop in the screeplot can be used to decide on the number of principal components to be used in the data analysis. From Fig. 5.3A, it is clear that the first two principal components describe over 90% of the data variance and are adequate to explain the dataset. The principal component loadings are the coefficients of the principal components and are

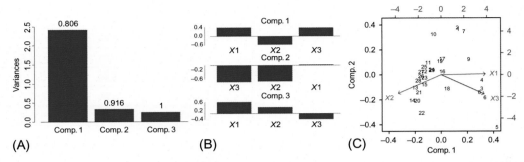

FIG. 5.3   **(A) Screeplot displaying the fraction of the variance explained by the principal components. (B) Principal component loadings displaying the relationship between principal components and the original variables. (C) A biplot displaying the original variables and the transformed variables in the principal component axes.**

shown in Fig. 5.3B. For example, we can see that there is a positive correlation between principal component 1 and variables $X1$ and $X3$ whereas there is a negative correlation with variable $X2$. Understanding such relationship is valuable in gaining a physical understanding of the principal components as we will see later in the field application. Another way to display this relationship is using the biplot in Fig. 5.3C. It displays the original variables and the transformed variables using the principal component axes. For example, the biplot clearly shows that there is very little dependence between principal component 2 and the variable $X1$. This can be verified from the loadings for principal component 2 shown in Fig. 5.3B.

5.3 CLUSTER ANALYSIS

The goal of cluster analysis is to partition a dataset into groups that are internally homogeneous and externally distinct (Davis, 1986; Kaufman and Rousseeuw, 1990; Johnson and Wichern, 1992). The classification is carried out on the basis of a measure of similarity or dissimilarity within and between the groups. Cluster analysis can be viewed as an unsupervised method of pattern recognition because the operation is typically not guided by a priori hypothesis or external models. Variable selection plays an important role, and different choices may result in drastically different results. For example, if the purpose of cluster analysis is to characterize a geologic formation by identifying electrofacies groups from a suite of well logs, then we should select well logs that are sensitive to lithology (Doveton and Prensky, 1992). Experience and user intervention can be critical in the proper interpretation of results of cluster analysis. In two or three dimensions, clusters can be visualized. With more than three dimensions, we need some kind of analytic assistance to reduce the dimensionality of the data without the significant loss of information. One such approach is the principal component analysis discussed above. Generally speaking, clustering algorithms fall into three different categories: partitioning or relocation, hierarchical, and model-based clustering algorithms.

5.3.1 k-Means Clustering

Partitioning or relocation methods require the user to specify the initial number of groups or clusters, and the algorithm iteratively reallocates observations between groups until a predefined convergence criterion is reached. Most clustering algorithms rely on some distance or dissimilarity measure between data points to classify them into groups (Mahalanobis, 1936). The simplest and most common measure is the Euclidean distance between data vectors x_1 and x_2: $d(x_1, x_2) = \|x_1 - x_2\|$. One of the popular relocation methods is the k-means algorithm. In this procedure, the user specifies k-groups as the number of clusters along with their initial centroid locations. A matrix of similarities is then computed between the n data points and the k-centroids, and each observation is assigned to the group with the closest centroid. A new centroid, or the multidimensional version of the mean, for each group is computed, and the process is repeated. With each iteration, the centroids are expected to move toward the actual centers of the local groups formed during the process. Group labels are assigned by minimizing the within-cluster sum-of-squares distances for the k-groups, that is,

$$\text{Minimize} \sum_{g=1}^{k} N_g \sum_{x \in C_g} d(x, \overline{x}_g)^2 \tag{5.4}$$

where N_g is the size of C_g, the gth cluster.

A stepwise illustration of the k-means clustering is shown in Fig. 5.4. It involves iterative refinement with two steps in each iteration: (i) an assignment step where each observation is assigned to the closest mean and (ii) an updating step that computes the new means to be the centroids of the observations in the cluster. The algorithm proceeds until the within-cluster sum-of-squares distance in Eq. (5.4) is minimized. The use of the least-squares minimization method is a disadvantage that makes k-means less resistant to outliers. Also, the memory requirements are quadratic in the number of observations. The method requires specifying the number of clusters in advance and the centroids of the clusters to get started. The results can be sensitive to this initial choice, and often, prior knowledge can be used to define the initial clusters. For clustering using diverse data types, it is necessary to normalize the variables for stable and consistent results. The major advantage of the k-means algorithm is its computational efficiency because the algorithm operates on relatively small dissimilarity matrices.

An Illustrative Example of **k-Means Clustering**

We continue the analysis of the dataset in Fig. 5.2. In Fig. 5.5A, the data are displayed using all three principal components. The results of cluster analysis using k-means with the first two principal components are shown in Fig. 5.5B. Two data clusters have been identified as indicated by the varying symbol sizes. A visual inspection and comparison with the data cloud show how cluster analysis can reveal the underlying data structure. A summary of the main results from the k-means clustering is given in Table 5.2.

5.3.2 Hierarchical Clustering

Hierarchical clustering methods proceed by stages producing a sequence of partitions, each corresponding to a different number of clusters (Mardia et al., 1979). These algorithms

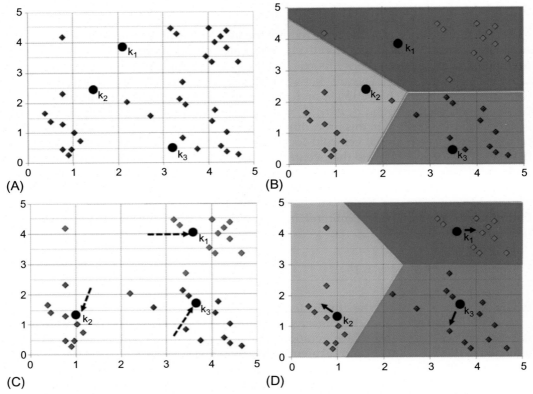

FIG. 5.4   (A) k initial "means" (in this case, $k = 3$) are randomly selected from the dataset. (B) k clusters are created by associating every observation with the nearest mean. (C) The centroid of each of the k clusters becomes the new means. (D) Steps 2–3 are repeated until convergence has been reached by minimizing the within-cluster sum of squares.

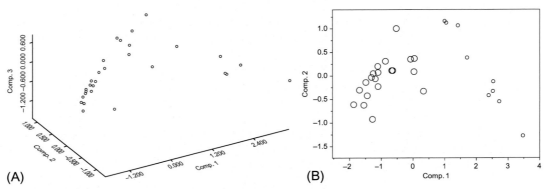

FIG. 5.5   (A) Data cloud represented using three principal components and (B) k-means clustering using first two principal components for $k = 2$.

TABLE 5.2 **Results of *k*-Means Clustering**

	Comp1	Comp2
Cluster centroids		
Cluster 1	2.09	0.114
Cluster 2	−0.941	−0.051
	Within-Cluster Sum of Squares	**Cluster Size**
Cluster size and sum of squares		
Cluster 1	11.62	9
Cluster 2	10.89	20

can be either agglomerative, meaning that groups are merged or divisive in which one or more groups are split at each stage. Hierarchical procedures describe a method yielding an entire hierarchy of clustering for the given dataset. The most common hierarchical algorithm is the agglomerative nesting, also known as Ward's method, which starts with each observation in separate groups and proceeds with the clustering until all observations are in a single group. Computation consists of iteratively merging two clusters with the smallest dissimilarity and then recomputing the dissimilarity between the new cluster and all the remaining clusters. The order and strength of the splits can be displayed using a classification tree or dendrogram. The dendrogram is a graphic representation of the entire hierarchical process and shows how the data are merged into clusters at various stages of the algorithm (Fig. 5.6). At the end of the agglomeration process, all the data points are merged into a single cluster, and typically, a distance or dissimilarity criterion is used to identify a natural breakpoint in the tree to identify the distinctive clusters. In practical applications, this may also be done based on inspection or some prior knowledge that can be used to validate the clusters.

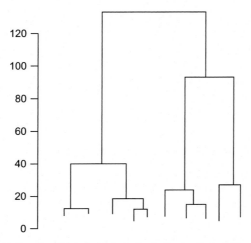

FIG. 5.6 **A dendrogram illustrating the clustering hierarchy in the agglomerative approach. A distance criterion is used to cut the tree and identify the clusters.**

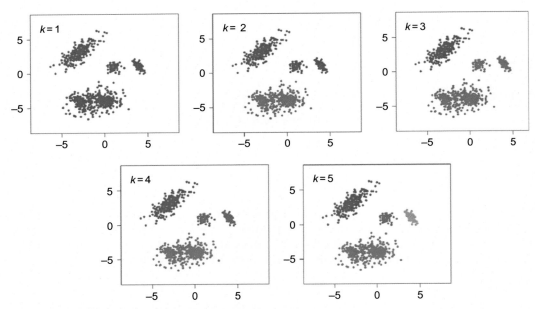

FIG. 5.7   **An illustration of the hierarchical clustering using the divisive approach.**

In the divisive method, all points begin in a single group (i.e., $k=1$), and the clusters are split apart as the number of groups k increases. Some of the criteria for splitting can be to choose the cluster with the largest diameter or to choose the cluster with largest within-cluster pairwise distance. A stepwise illustration of the procedure is shown in Fig. 5.7.

Unlike the k-means clustering, the hierarchical approach does not require specification of the number of clusters at the beginning, although it still requires the selection of clusters at the end based on some mathematical or external criterion. A major disadvantage of the approach is the computational cost when the number of data points is large. This requires manipulation of a potentially large similarity matrix. For example, for n observations, there will be $n(n-1)/2$ similarities, the cluster analysis will involve $(n-1)$ steps, and the computational cost tends to increase rapidly with the number of observations. To alleviate the situation, the cluster analysis is often carried out in conjunction with principal component analysis (PCA). The dominant principal components are used to carry out the clustering in a reduced dimensional space.

An Illustrative Example of Hierarchical Clustering

We now illustrate the hierarchical clustering using the sample dataset. Fig. 5.8 shows the dendrogram. As discussed before, it starts with each of the 29 data points as individual cluster, and these clusters are progressively merged, ultimately leading to a single cluster. A decision is then made about the number of clusters and a cutoff point is placed on the tree to assign the data to the clusters or groups. In Fig. 5.8, the cutoffs are shown for two clusters and three clusters. For the 2 cluster case, we can see that the dataset is grouped into clusters of 9 points and 20 points, the same outcome as the k-means clustering above.

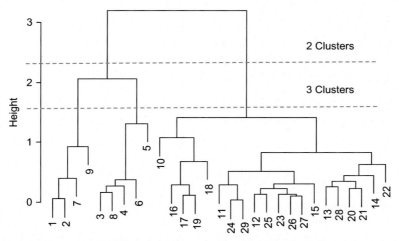

FIG. 5.8 **Dendrogram for the example dataset indicating cutoffs for two and three clusters.**

5.3.3 Model-Based Clustering

Neither hierarchical nor relocation methods directly address the issue of determining the number of groups within the data. However, various strategies for simultaneous determination of the number of clusters and cluster membership have been proposed, and one such method is the model-based clustering techniques such as the expectation-maximization (EM) algorithm. This approach can give much better performance than traditional procedures such as hierarchical and k-means clustering, which often fail to identify groups that are either of overlapping or of varying sizes and shapes. Another advantage of the model-based approach is that there is an associated Bayesian criterion for assessing the model. As discussed below, the method provides a means of selecting not only the parameterization of the model but also the number of clusters without the subjective judgments necessary in other conventional cluster analysis techniques.

The key idea of model-based clustering is that the data are generated by a mixture of underlying probability distributions. We assume that the probability density function of a p-dimensional observation $\mathbf{x}$ from the kth group is $f_k(\mathbf{x};\theta)$ for some unknown parameter vector θ. Given observations $\mathbf{D} = (\mathbf{x}_1, \ldots, \mathbf{x}_n)$, let $\boldsymbol{\gamma} = (\gamma_1, \ldots, \gamma_n)^{\mathrm{T}}$ denote the identifying group labels for the classification. The parameters θ and $\boldsymbol{\gamma}$ are determined so as to maximize the likelihood:

$$L(\mathbf{D}; \theta, \gamma) = \prod_{i=1}^{n} f_{\gamma i}(\mathbf{x}_i; \theta) \tag{5.5}$$

Note that $\gamma_t = k$ if $\mathbf{x}_i$ comes from the kth group.

In general, each cluster is represented by a multivariate Gaussian model:

$$f_c(\mathbf{x}_i | \boldsymbol{\mu}_c, \boldsymbol{\Sigma}_c) = (2\pi)^{-\frac{p}{2}} |\boldsymbol{\Sigma}_c|^{-1/2} \exp\left\{ -\frac{1}{2} (\mathbf{x}_i - \boldsymbol{\mu}_c)^T \sum_c^{-1} (\mathbf{x}_i - \boldsymbol{\mu}_c) \right\} \tag{5.6}$$

where $\mathbf{x}_i$ represents the data and c denotes an integer subscript specifying a particular cluster. Clusters are ellipsoidal, centered at the means $\boldsymbol{\mu}_c$. The covariances $\boldsymbol{\Sigma}_c$ determine their geometric characteristics. For example, the orientation of the clusters will be given by the eigenvectors of Σ_c, the cluster size or variance is given by the largest eigenvalue of Σ_c, and the ratio of the other eigenvalues to the largest determines the cluster shape. From Eq. (5.6), we can see that if we simplify the covariance matrix for each cluster to be diagonal and identical, $\Sigma_c = \sigma^2 I$ where I is the identity matrix, then maximizing the likelihood in Eq. (5.5) is the same as minimizing the sum-of-squares distances within the clusters and the algorithm reduces to the commonly used Ward's method discussed above. Thus, Ward's method assumes that the clusters are spherical in shape with the same size or variance.

5.4 DISCRIMINANT ANALYSIS

Discriminant analysis is a multivariate method for assigning an individual observation vector to two or more predefined groups on the basis of measurements. Unlike the cluster analysis, the discriminant analysis is a supervised technique and requires a training dataset with predefined groups. This technique is based on the assumption that an individual sample arises from one of g populations or groups $\Pi_1, \ldots, \Pi_g, g > 2$. If each group is characterized by a group-specific probability density or likelihood function $f_c(x)$ and the prior probability of the group π_c is known, then according to Bayes' theorem, the posterior distribution of the classes given the observation x is

$$p(c|x) = \frac{\pi_c p(x|c)}{p(x)} = \frac{\pi_c f_c(x)}{p(x)} \propto \pi_c f_c(x) \tag{5.7}$$

and the observations should be allocated into the group with the maximal posterior probability $p(c|x)$. Suppose the distribution for each group c be expressed by Eq. (5.6). Then, the Bayesian maximum a posteriori criterion is used to allocate a future observation x to the group c for which the function in Eq. (5.7) is the largest or its negative

$$\begin{aligned} Q_c &= -2\log f_c(x) - 2\log \pi_c \\ &= (x - \mu_c)^T \Sigma_c^{-1} (x - \mu_c) + \log|\Sigma_c| - 2\log \pi_c \end{aligned} \tag{5.8}$$

is the smallest. The first term of Eq. (5.8) is known as the *Mahalanobis distance* (Mahalanobis, 1936) from x to the group mean μ_c. The difference between the Q_c for the groups is a quadratic function of x, so the method is known as quadratic discriminant analysis and the boundaries of the decision regions are quadratic surfaces in x space. If the groups have a common covariance Σ, the differences in Q_c are then linear functions of x and create linear decision boundaries between groups. Then, this method reduces to linear discriminant analysis that assumes that the data groups have similar sizes and orientation.

The concept of linear discriminant analysis is best illustrated in Fig. 5.9 (Doveton and Prensky, 1992). In the figure, we can see that there is considerable overlap between the two data groups when seen as frequency curves on each axis. Given these data as the training data, assignment of groups to some test data becomes difficult, particularly for the overlapping regions. For this example, there is clear separation between groups A and B in the bivariate cross plot; however, in practice, most likely, there will be regions of overlap. The linear discriminant or allocation function is given by the equation of the line on which the distance between the data clouds is maximized while the spread within each cloud is

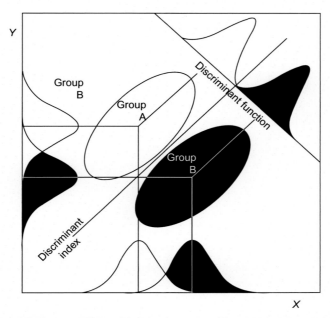

FIG. 5.9   **A graphic illustration of the discriminant function for a bivariate dataset.** *Reprinted with permission from Doveton, J.H., Prensky, S.E., 1992. Geological applications of wireline logs—a synopsis of developments and trends. Log Anal. 33 (1992) 286. Copyright, AAPG.*

minimized. This is illustrated in Fig. 5.9. All the data points can now be projected on this line. A discriminant index is computed as the midpoint of the projection of the mean of the data groups on this line. The discriminant index is then used as the boundary between the data groups to assign a new observation to one of the groups.

The discriminant analysis requires training data in the form of prior classification into relatively homogenous subgroups whose characteristics can be described by the statistical distributions of the grouping variables associated with each subgroup. Typically, the classification is performed by defining the distinct groups based on the unique characteristics of the measurements or by applying known external criteria. However, because in many situations a training dataset with absolutely known classifications is not easily obtained, a method like model-based cluster analysis is often used for classification purposes.

An Illustrative Example of Discriminant Analysis

We now illustrate the linear discriminant analysis using the sample dataset. The goal here is to identify a classification boundary for the two clusters identified by the k-means algorithm (Fig. 5.5) and develop a linear equation, called a discriminant function that best differentiates between the two groups. The classification boundary determined from linear discriminant analysis is shown in Fig. 5.10. Given a new set of $X1$, $X2$, and $X3$, we can now use the classification boundary or the discriminant function to assign each data to one of the two groups. Note that this is a supervised classification as it relies on the classification pattern established from the prior analysis.

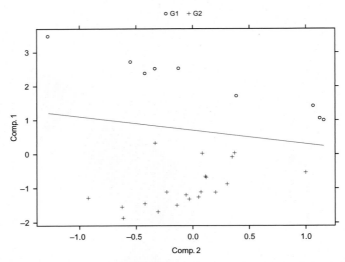

FIG. 5.10   **Classification boundary identified from linear discriminant analysis shown using the principal component axes.**

5.5 FIELD APPLICATION: THE SALT CREEK DATA SET

5.5.1 Dataset Description

We now demonstrate the application of the multivariate data analysis, specifically PCA, cluster analysis, and discriminant analysis using a field example (Lee et al., 2002). Our goal here is to predict permeability using a suite of well logs for the Salt Creek Field Unit (SCFU) data introduced in Chapter 4 (SALT-CREEK.DAT). The data used in this analysis belong to seven wells with cores and measured permeabilities for the cored interval. A suite of seven well logs (GR, LLD, MSFL, DT, NPHI, RHOB, and PEF) are used to predict permeability in this highly heterogeneous carbonate reservoir in the Permian Basin, Texas. Out of the seven cored wells, two (G517 and G520) were left out to verify the correlations using blind tests.

An important first step of data correlation is data partitioning whereby we subdivide the data into groups or classes that are internally homogeneous with respect to some predefined measure. A common approach for data partitioning is cluster analysis discussed before. For field applications with high-dimensional dataset, it is often necessary to perform the cluster analysis in a reduced dimensional space. The cluster analysis is typically carried out in the principal component space with the first few principal components explaining the majority of the data variance. This not only reduces the computational cost but also can help eliminate the effects of spurious noise in the data.

When the clusters are derived from well logs, they are often referred to as "electrofacies." So, electrofacies may be defined by a similar set of log responses that characterizes a specific rock type and allows it to be distinguished from others (Serra and Abbott, 1982). Once the well-log data are associated with a set of electrofacies, we can build correlation between measured permeability from cores and various well logs for each electrofacies. In this example, we will use the nonparametric regression method, namely, alternating conditional expectation (ACE) algorithm, discussed in Chapter 4 to build such correlation without any a priori assumption of functional form between permeability and well logs. Finally, the discriminant analysis is used to first assign electrofacies in "blind" wells and then predict permeability using appropriate correlation specific to the electrofacies. A stepwise illustration of the process is shown in Fig. 5.11. We now discuss below the steps involved in the analysis.

5.5.2 PCA

PCA is applied to obtain the principal components PC_j $(j=1,\ldots,7)$ from the well-log data after normalization. Fig. 5.12 shows the screeplot, a barplot of the variance of the principal components labeled by $\sum_{i=1}^{j} \lambda_i/\text{trace}(\mathbf{\Sigma})$, which often provides a convenient visual method of identifying the important principal components. The first four principal components in this case explain around 90% variation in the dataset.

In the scatterplot (Fig. 5.13), we explore the relationship between reservoir properties and the three major principal components generated from the seven well logs. The first principal component (PC1) appears to correlate with porosity of the formation, while the second principal component (PC2) shows a strong correlation with gamma-ray readings.

The eigenvectors of the correlation matrix provide coefficients or loadings of the principal component transformation (Table 5.3). For example, PC1 and PC2 are given by

$$PC1 = -0.12GR - 0.38\log(LLD) - 0.41\log(MSFL)$$
$$+ 0.47DT - 0.46RHOB + 0.48NPHI - 0.09PEF$$
$$PC2 = 0.63GR - 0.29\log(LLD) - 0.13\log(MSFL)$$
$$-0.14DT + 0.09RHOB - 0.09NPHI - 0.68PEF$$

We can now see that the principal components are simply surrogate variables defined by a weighted linear combination of the well logs.

5.5.3 Cluster Analysis

Model-based cluster analysis is used to define eight distinct groups based on the unique characteristics of the well-log measurements. In Fig. 5.14, each cluster can be treated as an electrofacies that reflects the petrophysical, lithologic, and diagenetic characteristics. The scatterplot in Fig. 5.13 relating the principal components to the physical variables can be useful in identifying the characteristics of the clusters or electrofacies. Based on the scatterplot, we can see that as PC1 increases, porosity decreases and the rocks become tighter. Also, as PC2 increases, the gamma-ray reading increases, and the rocks become more shaley. Thus, we can state qualitatively that the first electrofacies group (EF1) indicates tight media (low porosity) with low gamma-ray reading and the eighth electrofacies group (EF8) represents porous media with high gamma-ray reading.

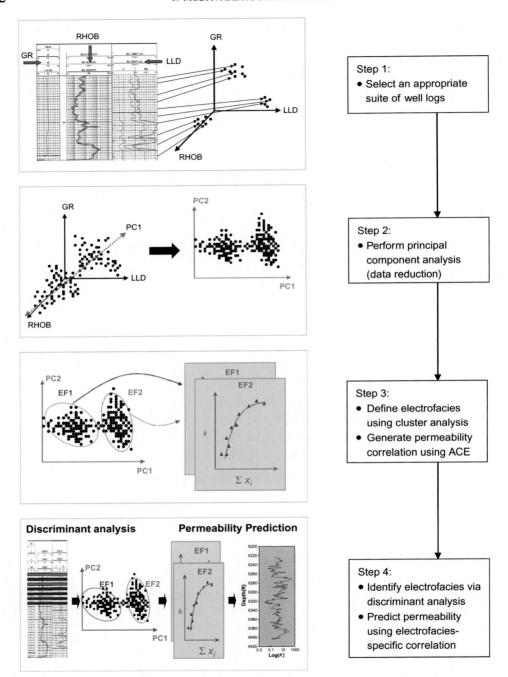

FIG. 5.11   **A schematic flowchart of permeability prediction based on electrofacies characterization.** *Reprinted from Lee, S.H., Khraghoria, A., Datta-Gupta, A., 2002. Electrofacies characterization and permeability predictions in carbonate reservoirs: role of multivariate analysis and non-parametric regression. SPE Reserv. Eval. Eng. 5(3). Copyright SPE.*

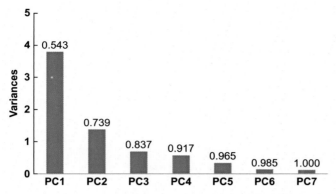

FIG. 5.12   Screeplot, a barplot labeled by the fraction of the total variance explained by the principal components. *Reprinted from Lee, S.H., Khraghoria, A., Datta-Gupta, A., 2002. Electrofacies characterization and permeability predictions in carbonate reservoirs: role of multivariate analysis and non-parametric regression. SPE Reserv. Eval. Eng. 5(3). Copyright SPE.*

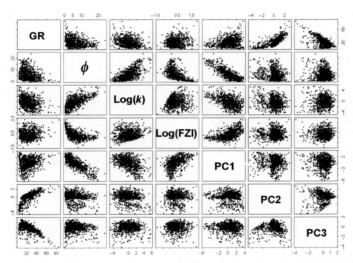

FIG. 5.13   Scatterplot of GR, core porosity, core permeability, flow zone indicator (FZI), PC1, PC2, and PC3 of well logs. *Reprinted from Lee, S.H., Khraghoria, A., Datta-Gupta, A., 2002. Electrofacies characterization and permeability predictions in carbonate reservoirs: role of multivariate analysis and non-parametric regression. SPE Reserv. Eval. Eng. 5(3). Copyright SPE.*

5.5.4 Data Correlation and Prediction

After partitioning of well-log responses into electrofacies groups, the nonparametric regression ACE algorithm is applied to model the correlation between permeability and well-log responses within each of the partitioned groups. Table 5.4 compares the regression errors for the model used for developing the correlations. These errors are summarized in terms of mean squared error (MSE) and mean absolute error (MAE) during regression.

TABLE 5.3   **Results of Principal Component Analysis of Well Logs**

	PC1	PC2	PC3	PC4	PC5	PC6	PC7
GR	−0.122	0.628	−0.761	−0.027	0.096	−0.015	0.033
Log(LLD)	−0.379	−0.293	−0.119	0.567	0.654	−0.072	−0.043
Log(MSFL)	−0.412	−0.127	−0.144	0.483	−0.742	−0.046	0.084
DT	0.470	−0.140	−0.198	0.161	−0.068	−0.805	−0.206
RHOB	−0.464	0.089	0.205	−0.340	0.079	−0.553	0.554
NPHI	0.476	−0.089	−0.124	0.283	0.040	0.172	0.799
PEF	−0.092	−0.684	−0.537	−0.472	−0.036	0.095	0.043
Eigenvalue	3.80	1.38	0.685	0.558	0.338	0.135	0.107
Contribution (%)	54.3	19.7	9.8	8.0	4.8	1.9	1.5
Cum. contribution (%)	54.3	74.0	83.7	91.7	96.5	98.5	100

Reprinted from Lee, S.H., Khraghoria, A., Datta-Gupta, A., 2002. Electrofacies characterization and permeability predictions in carbonate reservoirs: role of multivariate analysis and non-parametric regression. SPE Reserv. Eval. Eng. 5(3). Copyright SPE.

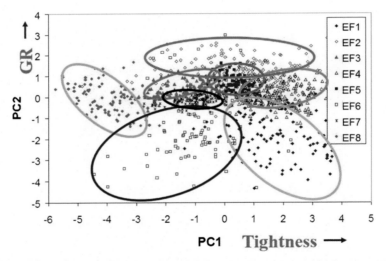

FIG. 5.14   **The distribution of electrofacies data plotted on the first two principal components of well logs.** *Reprinted from Lee, S.H., Khraghoria, A., Datta-Gupta, A., 2002. Electrofacies characterization and permeability predictions in carbonate reservoirs: role of multivariate analysis and non-parametric regression. SPE Reserv. Eval. Eng. 5(3). Copyright SPE.*

We now predict permeability in G517, one of the two cored wells that were left out for blind tests. The first step involves defining the electrofacies at various depths from the well-log responses in G517. On the basis of the eight clusters defined in step 2, an allocation function is determined by discriminant analysis. Through the allocation function, we can define the group membership of the log responses in these wells. Fig. 5.15 shows the electrofacies profile in well G517. Permeabilities are obtained by applying the correlation model derived in step 3 for each electrofacies, and the results are compared with measured data in Fig. 5.16A. Overall,

TABLE 5.4   **Comparison of Regression and Prediction Errors in Three Models**

	Error	ACE
Regression error (5 Wells, 904 sample pts.)	MAE	0.97
	MSE	1.58
Prediction error (G517, 174 sample pts.)	MAE	1.15
	MSE	2.25
Prediction error (G520, 183 sample pts.)	MAE	1.04
	MSE	1.74

Reprinted from Lee, S.H., Khraghoria, A., Datta-Gupta, A., 2002. Electrofacies characterization and permeability predictions in carbonate reservoirs: role of multivariate analysis and non-parametric regression. SPE Reserv. Eval. Eng. 5(3). Copyright SPE.

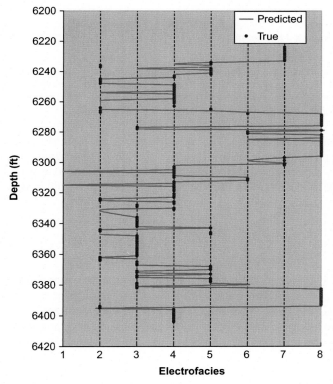

FIG. 5.15   **The profile of eight electrofacies at Well G517 determined by discriminant analysis. The "true" refers to electrofacies assigned when all the wells were included in the cluster analysis for electrofacies classification.** *Reprinted from Lee, S.H., Khraghoria, A., Datta-Gupta, A., 2002. Electrofacies characterization and permeability predictions in carbonate reservoirs: role of multivariate analysis and non-parametric regression. SPE Reserv. Eval. Eng. 5(3). Copyright SPE.*

the predicted permeabilities based on log data in G517 agree well with the core measurements. To illustrate the value of the multivariate analysis, specifically the data partitioning via electrofacies classification, we compare the permeability predictions with classification based on stratigraphic zonation (Fig. 5.16B). The electrofacies approach seems to significantly

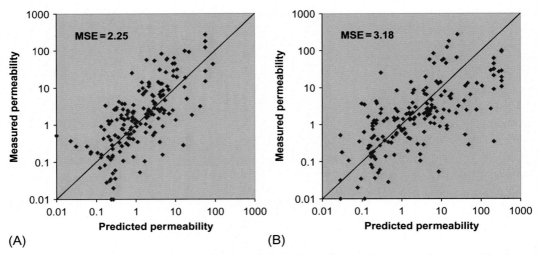

FIG. 5.16   Scatterplot of measured and predicted permeability based on (A) the electrofacies classification and ACE (Well G517) and (B) stratigraphic zonation. *Reprinted from Lee, S.H., Khraghoria, A., Datta-Gupta, A., 2002. Electrofacies characterization and permeability predictions in carbonate reservoirs: role of multivariate analysis and non-parametric regression. SPE Reserv. Eval. Eng. 5(3). Copyright SPE.*

outperform the zonation approach. This is to a large extent because of the improved discriminatory power of pattern recognition from the multivariate analysis (Lee et al., 2002).

The reader can reproduce the results in this field example using the software EFACIES and the Salt Creek field data (SALT-CREEK.DAT) made available in the online resources for this book.

5.6 SUMMARY

In this chapter, we have introduced three important techniques for multivariate data analysis, namely, the principal component analysis, cluster analysis, and discriminant analysis. We have illustrated the strength of the methods for data visualization, dimension reduction, understanding the intrinsic data structure, pattern recognition, and data partitioning/classification for regression analysis. The field application demonstrates the power and utility of multivariate analysis for improving data correlation and predictions.

Exercises

1. Use the dataset "MULTIVAR_FIG5-2.DAT" to perform the Principal Component Analysis as follows:

   (1) Make three plots: (1) x_1 vs. x_2, (2) x_2 vs. x_3, and (3) x_3 vs. x_1. Examine the correlation among the variables.

   (2) Compute the normalized (mean-zero and unit-variance) variables z_1, z_2, and z_3 corresponding to x_1, x_2, and x_3, respectively. (Hint: First, calculate the mean and

variance for the variables x_1, x_2, and x_3.) Then for each variable, subtract its mean and divide by its variance.

(3) Compute covariance matrix $C = Z^T Z/(n-1)$ where $Z = [z_1\ z_2\ z_3]$ and n is number of data.

(4) Perform a singular value decomposition of the covariance matrix ($C = Q^T \Lambda Q$) and compute the principal components. How many principal components would be necessary to represent the original three variables while preserving at least 90% of the variance in the data?

(5) Plot PC2 vs. PC1 and compute the variances along each axis. You will see that PC1 has much larger variance than PC2 as indicated by the eigenvalues.

2. Using the principal components 1 and 2 in Example 1, perform k-mean clustering and divide the data into two clusters. (Hint: Follow the steps in Fig. 5.4 by first randomly selecting two cluster means and then assigning the principal components to the cluster with the nearest mean. Compute the coordinates of two cluster centroids by averaging PC1 and PC2 in each cluster. Update the cluster number based on the new distance. Repeat this procedure until the convergence is achieved.)

3. Using the dataset "SALT-CREEK.DAT," build permeability prediction models using well logs and nonparametric regression.

(1) Perform the PCA and reproduce the results in Fig. 5.12 and Table 5.3.

(2) Using the principal components 1 and 2, perform the k-mean clustering with three different cluster numbers. Suggest the appropriate number of clusters by examining the histograms of the well logs within each cluster.

(3) Build permeability prediction models for each cluster using ACE algorithm.

(4) Predict the electrofacies and permeability for the blind well 517 (SALT-CREEK-G517.DAT).

(5) Compare the R^2-value between predicted and measured permeability with the Example 6 in Chapter 4 where the electrofacies was not accounted in the regression. Does electrofacies classification improve the prediction quality?

4. Using the dataset Multivar_Exercise.xlsx, complete the following principal component analysis:

(1) Calculate the data correlation matrix for five independent variables (x_1 to x_5).

(2) Compute the eigenvalues and eigenvectors of the correlation matrix.

(3) Provide a screeplot for the variance analysis. How many principal components are sufficient to explain 90% percentage of the variance?

5. Using the dataset Multivar_Exercise.xlsx, perform k-mean clustering of the five independent variables (x_1 to x_5). (Specify $k = 3$ for the number of clusters.)

6. Using the dataset Multivar_Exercise.xlsx, draw scatterplot matrix, which is similar to Fig. 5.13, for the following:

(1) For all five independent variables (x_1 to x_5) to examine any correlation among them.

(2) For the first two principal components with different colors or symbols representing the three clusters from k-mean clustering.

(3) For the first two principal components with different colors or symbols representing the dependent variable y, from the given data.

(4) Please comment on the results of cluster analysis based on the visual examinations.

References

Davis, J.C., 1986. Statistics and Data Analysis in Geology, second ed. John Wiley & Son, New York. p. 527.

Doveton, J.H., Prensky, S.E., 1992. Geological applications of wireline logs—a synopsis of developments and trends. Log Anal. 33, 286.

Hastie, T., Tibshirani, R., Friedman, J.H., 2008. The Elements of Statistical Learning: Data Mining, Inference, and Prediction, second ed. Springer, New York.

Hempkins, W.B., 1978. Multivariate Statistical Analysis in Formation Evaluation. Society of Petroleum Engineers. https://doi.org/10.2118/7144-MS.

Hempkins, W.B., Kingsborough, R.H., Lohec, W.E., Nini, C.J., 1987. Multivariate Statistical Analysis of Stuck Drillpipe Situations. Society of Petroleum Engineers. https://doi.org/10.2118/14181-PA.

Johnson, R.A., Wichern, D.W., 1992. Applied Multivariate Statistical Analysis. vol. 4. Prentice hall, Englewood Cliffs, NJ.

Kaufman, L., Rousseeuw, P., 1990. Finding Groups in Data: An Introduction to Cluster Analysis. Wiley, New York.

Lee, S.H., Khraghoria, A., Datta-Gupta, A., 2002. Electrofacies characterization and permeability predictions in carbonate reservoirs: role of multivariate analysis and non-parametric regression. SPE Reserv. Eval. Eng. 5 (3), 237–248.

Mahalanobis, P.C., 1936. On generalized distance in statistics. Proc. Natl. Inst. Sci. India 12, 49.

Mardia, K.V., Kent, J.T., Bibby, J.M., 1979. Multivariate Analysis. Academic Press, London. p. 521.

Mwenifumbo, C.J., 1993. Kernel Density Estimation in the Analysis and Presentation of Borehole Geophysical Data. Society of Petrophysicists and Well-Log Analysts.

Nitters, G., Davies, D.R., Epping, W.J.M., 1995. Discriminant Analysis and Neural Nets: Valuable Tools to Optimize Completion Practices. Society of Petroleum Engineers. https://doi.org/10.2118/21699-PA.

Scheevel, J.R., Payrazyan, K., 2001. Principal Component Analysis Applied to 3D Seismic Data for Reservoir Property Estimation. Society of Petroleum Engineers. https://doi.org/10.2118/69739-PA.

Serra, O., Abbott, H.T., 1982. The contribution of logging data to sedimentology and stratigraphy. SPEJ, 117–131. https://doi.org/10.2118/9270-PA.

Siena, M., Guadagnini, A., Della Rossa, E., Lamberti, A., Masserano, F., Rotondi, M., 2016. A Novel Enhanced-Oil-Recovery Screening Approach Based on Bayesian Clustering and Principal-Component Analysis. Society of Petroleum Engineers. https://doi.org/10.2118/174315-PA.

Strang, G., 1998. Introduction to Linear Algebra, third ed. Wellesley-Cambridge Press, Wellesley, MA, ISBN: 0-9614088-5-5.

Further Reading

Banfield, J.D., Raftery, A.E., 1993. Model-based Gaussian and non-Gaussian clustering. Biometrics 49, 803.

Eto, K., Suzuki, S., Samizo, N., Ichikawa, M. Electrical Facies: The Key to the Carbonate Reservoir Characterization. Personal Communications.

James, G., Witten, D., Hastie, T., Tibshirani, R., 2013. An Introduction to Statistical Learning. Springer, New York Vol. 112.

6

Uncertainty Quantification

The focus of this chapter is uncertainty quantification, which involves translating the uncertainty in the inputs of a model into the corresponding uncertainty in model outputs. To this end, we present a systematic approach regarding how to characterize the uncertainties, propagate them through the system model of interest into uncertainties in model predictions, and analyze the relative importance of various sources of uncertainty.

6.1 INTRODUCTION

6.1.1 Deterministic Versus Probabilistic Approach

Petroleum engineers and geoscientists dealing with the movement of fluids in subsurface geologic systems are often confronted with uncertainty caused by incomplete knowledge (arising from data gaps, measurement error, lack of resolution, biased sampling, etc.) and/or natural randomness. Examples of such geosystems include petroleum reservoirs and groundwater-bearing formations, as well as potential host rocks for carbon sequestration, natural gas storage, and nuclear waste disposal. The certainty and ubiquity of uncertainty in geosystems pose interesting challenges in the analysis and modeling of such systems. Traditional deterministic modeling of uncertainty has involved using best-guess or worst-case assumptions about model inputs to quantify their impacts on model predictions. Alternatively, a set of optimistic and pessimistic values are sometimes utilized to provide upside and downside forecasts around a reference scenario (Ovreberg et al., 1992). However, this simplistic approach is not capable of dealing with complex problems where the system response is nonlinear or where correlations exist between model parameters. Systematic combinations of optimistic and pessimistic values may also lead to confidence intervals that are too wide (resulting in overdesign) and whose reliability is difficult to assess.

Recently, there has been renewed interest in the use of probabilistic techniques for formal uncertainty quantification in petroleum reservoirs (e.g., Murtha, 1994; Bratvold and Begg, 2010; Ma and LaPointe, 2010; Caers, 2011), building upon the earlier work by several researchers (e.g., Walstrom et al., 1967; MacDonald and Campbell, 1986). In the probabilistic approach, multiple values of model parameters (taken from parameter-value distributions) are propagated through the system model to produce multiple valued consequences (or output distributions). Compared with the deterministic approach, the probabilistic approach offers multiple advantages. First, it allows all information regarding uncertain and variable parameters to be captured—as compared with using only the best-guess or worst-case values in the deterministic approach. Second, the full range of possible outcomes (as well as the probability of each outcome) can be quantified, whereas the deterministic approach cannot provide the likelihood associated with the outcome from any combination of scenarios. Third, it is possible to analyze the relationship between inputs and outputs to identify critical uncertain inputs while considering any synergistic effects between the model inputs. Thus, the probabilistic approach is better suited for making informed decisions under uncertainty.

As an illustrative example, consider the problem of predicting the uncertainty in future income from a new oil-producing reservoir. The uncertain variables are oil production, capital and operating expenses, and oil price. The uncertainty in each variable is characterized as (best-guess value)$\pm 10\%$. The goal of this exercise is to quantify the impact of these

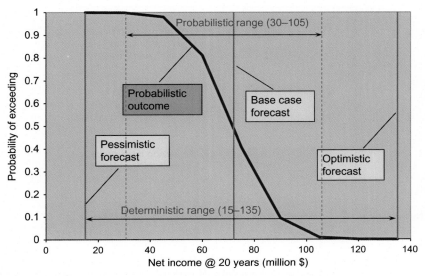

FIG. 6.1   **Example showing deterministic versus probabilistic analysis results.**

uncertainties via deterministic range analysis and probabilistic analysis. Fig. 6.1 summarizes the results of this analysis. Here, the reference case forecast for the deterministic analysis is based on combining all of the "best-guess" values. Similarly, the pessimistic and optimistic forecasts are derived from a systematic combination of all pessimistic or all optimistic values, respectively. The corresponding deterministic range is found to be between 15 and 135 million dollars, without any additional information regarding the relative likelihood of these or intermediate values. On the other hand, the probabilistic analysis does provide such information as shown by the probabilistic outcome curve. It is also clear that the deterministic analysis includes very low-probability outcomes at both extremes, which is why the probabilistic range (30–105 million dollars) is smaller and potentially more realistic. Although not shown here, the probabilistic analysis also reveals that the uncertainty in oil rate and oil price contribute the most to the overall spread in predictions of net income.

6.1.2 Elements of a Systematic Framework

As a first step in modeling under uncertainty, it is useful to outline the elements of a systematic framework for uncertainty quantification (Mishra, 2009). As schematically shown in Fig. 6.2, these are:

- *Uncertainty characterization*—which involves capturing all information regarding uncertain and variable factors by fitting and/or assigning marginal and joint distributions to uncertain model inputs
- *Uncertainty propagation*—which involves quantifying the full range of possible outcomes and the probability of each outcome by mapping the uncertainty in model inputs into the corresponding uncertainty in model outputs

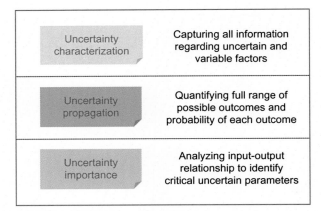

FIG. 6.2   **Key elements of a systematic framework for uncertainty quantification.**

- *Uncertainty importance assessment*—which involves analyzing input-output
  relationships to determine the key drivers of output uncertainty (aka "heavy hitters")

It should be noted that in the subsurface modeling and analysis literature, the term *"uncertainty analysis"* is commonly used to describe both uncertainty characterization and uncertainty propagation as defined above, whereas the term *"sensitivity analysis"* is used to describe the assessment of uncertainty importance (e.g., Ma and LaPointe, 2010). In other words, uncertainty analysis refers to the process of capturing all information regarding uncertain and variable factors and estimating distributions around model predictions. Sensitivity analysis, on the other hand, involves identifying key input parameters that contribute the most to the model's predictive uncertainty. In this book, we prefer to use the three descriptors of Fig. 6.2 to describe the broader uncertainty quantification process.

6.1.3 Role of Monte Carlo Simulation

A review of the petroleum and environmental geosciences literature shows that uncertainty analysis is often taken to be synonymous with Monte Carlo simulation (MCS). MCS can be broadly described as a numerical method for solving problems by random sampling (Morgan and Henrion, 1990).

As shown in Fig. 6.3, the probabilistic modeling approach of MCS allows a full mapping of the uncertainty in model parameters (inputs) and future system states (scenarios), expressed as probability distributions, into the corresponding uncertainty in model predictions (output), which is also expressed in terms of a probability distribution. Uncertainty in the model outcome is quantified via multiple model calculations using parameter values and future states drawn randomly from prescribed probability distributions. MCS is also known as the method of statistical trials because it uses multiple realizations (i.e., combinations of values) of different inputs to compute a probabilistic outcome.

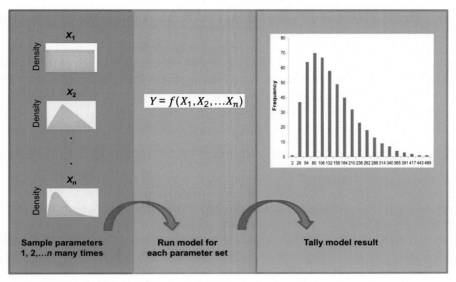

FIG. 6.3   **Schematic of the Monte Carlo simulation process.**

The MCS approach is computationally burdensome because it typically requires several hundred model calculations (if not more). However, as noted earlier, the underlying probabilistic framework also provides important information not available from a deterministic "best-guess" or "worst-case" calculation. It should also be noted that although MCS provides the greatest versatility in uncertainty propagation studies, it may not be the most efficient when (a) parameter uncertainty is poorly defined, (b) forward models are computation intensive, or (c) outcomes of interest are limited in number (Mishra, 1998). We discuss this further in Section 6.5.

In this chapter, we present a systematic approach for uncertainty quantification—primarily within a MCS framework—organized along the three elements shown in Fig. 6.2. The motivation here is to provide sufficient information to emphasize that there is more to MCS than simple random sampling, multiple model runs, and aggregation of results. Our MCS workflow can be described as follows:

(a) *Uncertainty characterization*
   (1) Select imprecisely known model input parameters to be sampled.
   (2) Construct probability distribution functions for each of these parameters.
(b) *Uncertainty propagation*
   (1) Generate a sample set by selecting a parameter value from each distribution.
   (2) Calculate the model outcome for each sample set and aggregate results for all samples (equally likely parameter sets).
(c) *Uncertainty importance*
   (1) Analyze the probabilistic calculations to determine the input-output relationships.
   (2) Identify the key uncertain parameters.

6.2 UNCERTAINTY CHARACTERIZATION

In this section, we describe how to select imprecisely known model input parameters to be sampled and how to assign ranges and probability distributions for each of these parameters. We will also discuss the problem of scale and how it affects the choice of distribution assignment. This is particularly important for parameters associated with spatially averaged models (e.g., volumetric reserves estimation).

6.2.1 Screening for Key Uncertain Inputs

As a first step in uncertainty characterization, it is useful to consider the selection of key uncertain inputs so as to identify and retain only those input variables that have the greatest impact on the outcomes of interest. Eliminating redundant uncertain inputs from the sampled set generally helps focus data collection efforts and improves the stability and reliability of MCS results. It also facilitates robust statistical model building of input-output relationships during the sensitivity analysis phase needed to identify key drivers of output uncertainty. This is most readily carried out using standard one-parameter-at-a-time (OPAT) sensitivity analysis with the results plotted using a *spider chart* or a *tornado chart*.

As an example of a spider chart, Fig. 6.4 shows the results of an OPAT sensitivity analysis for a problem involving CO_2 injection into a brine-filled formation (Ravi Ganesh and Mishra, 2016). Along the x-axis, the values of different variables are represented as indicators (e.g., $-1 =$ low, $0 =$ reference, and $+1 =$ high), which allows variables with different units to be effectively normalized. Sometimes, it is useful to vary all parameters over a common range (i.e., ± 2 standard deviations from the mean). Along the y-axis, either the unadjusted response or the change in the response from the reference case is plotted. The steeper the slope of the corresponding line for any variable, the greater its influence. Also, the curvature of each

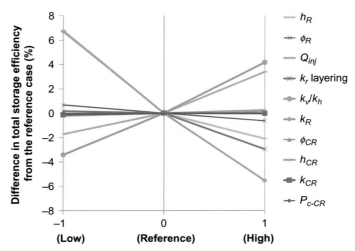

FIG. 6.4   **Example spider chart from OPAT sensitivity analysis for CO_2 injection into a brine-filled formation.** *After Ravi Ganesh, P., Mishra, S., 2016. Simplified physics model of CO2 plume extent in stratified aquifer-caprock systems. Greenhouse Gas Sci. Technol. 6, 70–82. https://doi.org/10.1002/ghg.1537.*

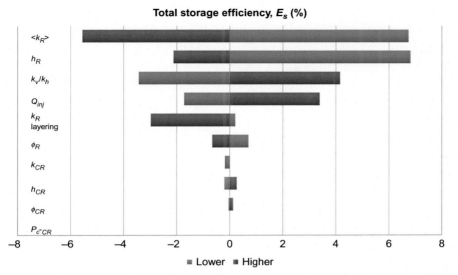

FIG. 6.5 **Example tornado chart from OPAT sensitivity analysis for CO_2 injection into a brine-filled formation.** *After Ravi Ganesh, P., Mishra, S., 2016. Simplified physics model of CO2 plume extent in stratified aquifer-caprock systems. Greenhouse Gas Sci. Technol. 6, 70–82. https://doi.org/10.1002/ghg.1537.*

line is indicative of model nonlinearity. In this specific case, the most sensitive parameter affecting total storage efficiency can be identified as mean reservoir permeability (k_R), permeability anisotropy ratio (k_v/k_h), nature of permeability layering (k_R layering), and CO_2 injection rate (Q_{inj}). Therefore, these should be treated as uncertainties in an MCS exercise, while the other parameters are fixed at their mean or median values.

A tornado chart captures sensitivities from a range analysis, where one starts with ranges of interest for the parameters (e.g., minimum and maximum, 5th and 95th percentiles). The model is run for each extreme value for each parameter, while keeping everything else fixed at their nominal values (i.e., mean and median). The resulting data are plotted as horizontal bars that show the full range of model output response for each of the uncertain inputs. The data are arranged such that the biggest response (i.e., widest bar) is plotted at the top, leading to the shape of a tornado. An example of such a plot is given in Fig. 6.5, which essentially confirms the results of the spider chart analysis. The main difference is that the tornado chart can only reflect the two end-member outcomes, whereas a spider chart can be populated with intermediate results by calculating model outcomes at values in between the reference case and the high/low points.

Another approach involves the use of experimental design-based screening techniques such as Plackett-Burman (PB) analysis. As described in Section 7.2.1, in this approach, a two-level design (i.e., high and low values) is used to estimate the main effects of the predictors on the response. Arinkoola and Ogbe (2015) present a case study showing how this approach is used to identify the "heavy hitters" for an uncertainty assessment of cumulative oil production forecasts from a reservoir model.

Once the key uncertain inputs have been identified, their proper characterization using probability distributions is an important step in producing a defensible uncertainty analysis study. Unfortunately, a systematic approach to probability distribution assignment often appears to be ignored in the petroleum geoscience literature. Such a methodology has been described in detail by Mishra (2002) and involves the following components:

- Fitting distributions to measured data using probability plotting or parameter estimation techniques (e.g., Hahn and Shapiro, 1967).
- Deriving distributions using known constraints and the principle of maximum entropy (e.g., Harr, 1987), which forces the analyst to be maximally uncertain with respect to unknown information.
- Assessing subjective distributions using formal expert elicitation protocols (e.g., Keeny and von Winterfeld, 1991).

6.2.2 Fitting Distributions to Data

As noted earlier in Section 3.4 only a handful of distributions are generally considered in practice (see Table 3.2). For example, uniform or triangular distributions are useful for representing low state of knowledge and/or subjective judgment, normal or lognormal distributions are commonly used to model errors due to additive or multiplicative processes, beta distributions are used for representing bounded, unimodal random variables, and Weibull distributions are popular for modeling component failure rates.

For normal or lognormal distributions, probability plotting is a convenient way for comparing the data with the postulated distribution and estimating its parameters. As described in Section 3.4.1, a probability plot for a normal (or lognormal) distribution is a graph of the ranked observation, x_i, (or $\ln x_i$) versus an approximation of the expected value of the inverse normal CDF, $G^{-1}(q_i)$. Recall that q_i is the quantile (cumulative frequency) of the empirical distribution, generally calculated using the Weibull formula, $q_i = i/(N+1)$ where i is the rank of the observation (sorted from smallest to largest) and N is the number of observations. Also, the inverse normal CDF, or the z-score, can be readily calculated using the Microsoft Excel function, NORMSINV.

A more flexible approach, which works for any distribution, involves the use of nonlinear least-squares analysis, as described in Section 3.4.2. Here, the goal is to estimate model parameters by minimizing the mean squared difference between the observed and predicted CDF. This process can be readily implemented using the nonlinear optimization package SOLVER in Microsoft Excel. Initial guesses for the parameter estimation process can be generated using the method of moments.

EXAMPLE 6.1 Fitting a lognormal distribution to observed data

Permeability values from multiple core samples in a well (PERM_FIG6-6.DAT) were found to be $x = (2.5, 4.2, 8.2, 10.1, 13.1, 14.7, 21.4, 24.2, 28.0, 32.2, 38.4, 44.5, 54.9, 72.3, 109, 221)$ mD. Fit these data to a lognormal distribution, and calculate the parameters of the distribution.

Solution

A lognormal distribution was first fit to the data using the probability plotting method. This requires plotting the natural logarithm of x against the inverse of the standard normal CDF, $G^{-1}(q)$, where q is the quantile. As shown in Fig. 6.6A, a very good fit was obtained except at the extreme

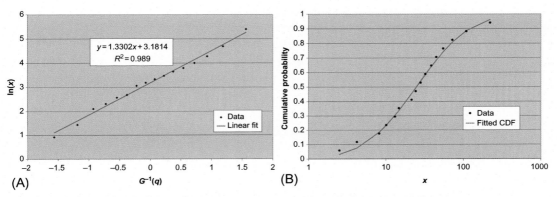

FIG. 6.6 **Lognormal fit using (A) probability plot and (B) nonlinear regression, Example 6.1.**

tails, with an R^2 value approximately equal to 1. The lognormal parameters are calculated from the slope and intercept of the best-fit line on the probability plot as $\alpha = 3.18$ and $\beta = 1.33$.

Next, these parameters are obtained using nonlinear least-squares analysis, which requires minimizing the sum of the squared differences between the observed and the predicted quantiles corresponding to each observed value. The Excel function NORMSDIST was used to generate the standard normal CDF necessary for estimating the cumulative probability. The corresponding best-fit parameters, obtained using the SOLVER toolbox in Excel, are $\alpha = 3.21$ and $\beta = 1.23$, which agree very well with those estimated using the probability plotting method. Fig. 6.6B compares the observed CDF with the predictions using regression parameters.

6.2.3 Maximum Entropy Distribution Selection

Although it is desirable to generate probability distributions for uncertain parameters on the basis of observed and/or simulated data, reality does not always cooperate with the analyst in this regard. Distributions are therefore routinely inferred on the basis of only a limited amount of information and are also subject to rather ad hoc assumptions. As an alternative, the principle of maximum entropy offers a systematic approach to distribution selection under such conditions.

It is well-known that the concept of thermodynamic entropy is related to the degree of disorder. Similarly, the concept of "information" entropy (Shannon, 1948) may be used to characterize the uncertainty of probability states, namely,

$$H = -\sum_i p_i \ln(p_i) \tag{6.1}$$

where H is the Shannon entropy (so named after its original proponent) and p_i is the probability associated with the ith sample. It is easily shown that the maximum entropy corresponds to a uniform distribution, where all samples are equally likely (Harr, 1987).

TABLE 6.1   **Maximum Entropy Distributions**

Constraint	Assigned PDF
Upper bound, lower bound	Uniform
Minimum, maximum, mode	Triangular
Mean, standard deviation	Normal
Range, mean, standard deviation	Beta
Mean occurrence rate	Poisson

Any other distribution would have a concentration of probability away from the extreme values, leading to a reduction of uncertainty and hence a reduction of entropy.

The principle of maximum entropy seeks to choose a PDF that maximizes the entropy, subject to known constraints. Uncertainty is reduced as much as possible by using all information (i.e., satisfying all constraints), but no further by unnecessary assumptions. This ensures that ignorance is preserved and one is maximally uncertain with respect to the unknown information. Harr (1987) discusses how the maximum entropy principle can help assign probability distributions on the basis of known constraints, as summarized in Table 6.1.

As an example, consider the situation when only the lower and upper bounds for porosity values are estimated to be 8% and 18%, based on a limited amount of data from the formation of interest and data from an analog formation. The principle of maximum entropy would indicate a uniform distribution for this case. One could opt for a triangular distribution, where the mode is taken as the midpoint of the range (i.e., 13%). However, that would be tantamount to making assumptions not supported by the data regarding the symmetry (or lack thereof) in the distribution. If the most likely value is known with some degree of certainty (i.e., 15%), then only should one resort to a triangular distribution. Thus, the entropy-based distribution selection framework forces the analyst to be maximally uncertain about the data.

6.2.4 Generation of Subjective Probability Distributions

Another common strategy employed in the absence of data is to ask subject matter experts to develop distributions representing their degree of belief regarding the uncertain quantity of interest. It is generally recommended (Helton, 1993) that distributions are best developed by specifying selected percentile values, rather than trying to specify a particular parametric distribution model (e.g., normal) and its associated parameters (e.g., mean and standard deviation).

In practice, one starts by specifying the minimum, the maximum, and the median values—which correspond to the 0th, 100th, and 50th percentiles. The distribution is refined by adding intermediate percentiles such as the 10th and 90th and the 25th and 75th. Plotting the empirical CDF also helps in deciding whether the selected values at given percentiles need to be adjusted, and/or additional percentiles need to be added. In general, it is easier for experts to defend the choice of values corresponding to selected percentiles than the choice of parameters characterizing a parametric distribution model. One helpful tool in this context is the probability scale used by the Intergovernmental Panel for Climate Change (IPCC, 2010), as shown in Table 6.2.

TABLE 6.2   IPCC Probability Scale for Subjective Assessments

Subjective Descriptor	Equivalent Cumulative Probability
Virtually certain	>0.99
Very likely	0.90–0.99
Likely	0.66–0.90
About as likely as not	0.33–0.66
Unlikely	0.10–0.33
Very unlikely	0.01–0.10
Exceptionally unlikely	<0.01

TABLE 6.3   Example Subjective Probability Assignment

Percentile	Expert 1	Expert 2	Consensus
0 (minimum)	0	2	1
10	1.5	2.5	2
25	3	3	3
50	4	4	4
75	4.3	6	6
90	4.7	7.5	7
100	5	9	8

As an illustrative example, consider the subjective assessment of values for well skin to be used in a history matching exercise. CDFs elicited from two experts are shown in Table 6.3. The first expert clearly has a preference for low skin values (mean = 3.9 and median = 4), whereas the second expert prefers high skin values (mean = 5.2 and median = 4). After much discussion, a consensus CDF was chosen that was more weighted toward lower values below the median and more weighted toward the higher values above the median, as shown below. It should be noted that although formal expert elicitation protocols as described above are not commonly employed in the petroleum geosciences, they do provide a traceable and defensible approach to assigning distributions based on subjective judgment.

When many uncertain quantities are candidates for subjective probability distributions, it is not worthwhile spending limited resources to develop such distributions for each and every parameter. Helton (1993) suggests a two-step procedure, wherein all variables are first crudely characterized as uniform (or log-uniform, depending on the range) distributions for a screening-level analysis. Model results are analyzed to identify the most important contributors to output uncertainty. Resources can then be focused on this subset of parameters for a more detailed characterization of uncertainty prior to the second-level analysis. Surrogate models, built using an experimental design approach (e.g., Chapter 7), can be valuable tools for this purpose.

6.2.5 Problem of Scale

Another important consideration in the assignment of proper distributions is the problem of scale. In petroleum and environmental geosciences, there is often a disparity between the data collection scale and the model discretization scale. As shown in Fig. 6.7, the former (left panel) is the scale associated with the physical variable, which is typically $\sim 10^{-2}$ m and corresponds to a higher variance, whereas the latter (right panel) is the scale associated with the model parameter, which is typically $\sim 10^{1}$ m and corresponds to a lower variance.

Since most model parameters are typically spatially averaged quantities, care must be taken to relate uncertainty/variability at the data collection scale to that required by the model. In particular, it should be noted that the variance observed at the data collection scale (e.g., core samples) reflects spatial variability and is much larger than that applicable at the scale of the model parameter (e.g., grid block average) because of spatial averaging.

As a simple example, consider the 10-sample net pay data from Example 3.9: h (ft) $= (13, 17, 15, 23, 27, 29, 18, 27, 20,$ and $24)$. It is required to develop a distribution for the *average* net pay to be used in a probabilistic reserves calculation. What is therefore needed is a characterization of the uncertainty around the average value, *not the full distribution of net pay itself*. As shown in Example 3.9, this is a t distribution with nine degrees of freedom. The sample mean (X) is 21.3, and sample standard deviation (s) is 5.52, from which standard error of mean (SE) $= 5.52/\sqrt{10} = 1.75$. For practical purposes, unless the number of samples is extremely small (i.e., < 5), we could also approximate this as a normal distribution with the same mean and standard error of mean.

The calculation of the CDF of the mean using the t distribution and its approximation as a normal distribution are shown below in Table 6.4. Here, the t- and z-values corresponding to a given percentile (and the degrees of freedom, if needed) are generated using the Excel

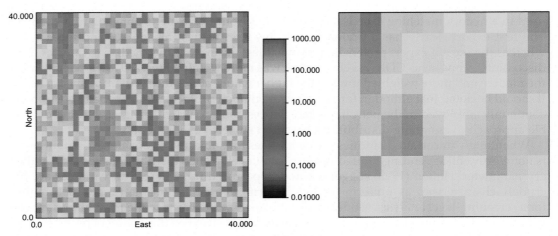

FIG. 6.7 **Example showing disparity between data collection scale for physical variables (left) and discretization scale for model parameters (right).**

TABLE 6.4   Example Calculation of CDF for Mean Using t and Normal Distributions

			n Number of Samples	=10			
			X_bar Sample Mean	=21.3			
			SE Standard Error of Mean	=1.75			

Net Pay (ft)	Rank	Quantile	Percentile	t-Value	t-Dist Calculated Net Pay (ft)	z-Value	Normal Calculated Net Pay (ft)
13	1	0.091					
15	2	0.182	0.01	−2.82	16.4	−2.33	17.2
17	3	0.273	0.05	−1.83	18.1	−1.64	18.4
18	4	0.364	0.1	−1.38	18.9	−1.28	19.1
20	5	0.455	0.25	−0.70	20.1	−0.67	20.1
23	6	0.545	0.5	0.00	21.3	0.00	21.3
24	7	0.636	0.75	0.70	22.5	0.67	22.5
27	8	0.727	0.9	1.38	23.7	1.28	23.5
27	9	0.818	0.99	2.82	26.2	2.33	25.4
29	10	0.909					

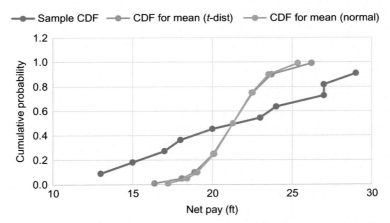

FIG. 6.8   Comparison of CDFs for a variable itself and its mean—using data from Example 3.9.

functions T.INV and NORMSINV, respectively. These standardized values are then used with the sample mean and the standard error of the mean to calculate the values of net pay corresponding to any given percentile.

Fig. 6.8 shows the CDF of the sample net pay data, along with a CDF computed for the mean using the t distribution, and also approximated using the normal distribution. The main takeaways from this graph are as follows: (a) The distribution for the mean is much tighter than that for the variable itself, and (b) the normal approximation to the t distribution works reasonably well even with as few as 10 samples.

6.3 UNCERTAINTY PROPAGATION

In this section, we describe how to translate the uncertainty in model inputs into the corresponding uncertainties in model predictions so as to capture (a) the full range of possible outcomes and (b) the probability of each outcome. This involves generating many sample sets (realizations) with random values of model parameters, selecting the appropriate number of runs needed to reliably estimate uncertainty in model outcomes, and approaches for presenting the MCS results. We also present some alternative uncertainty analysis techniques that complement MCS.

6.3.1 Sampling Methods

Random Sampling

The basic idea behind the random sampling approach is as follows. Let the CDF for any random variable, X, be denoted as $F_x(x)$, which is a nondecreasing function of x such that $0 \leq F_x(x) \leq 1$. Thus, we can establish a unique functional relationship $F_x^{-1}(u)$ for any u so long as $0 \leq u \leq 1$ and $F_x(x)$ is strictly monotonic, which is generally the case with probability distributions encountered in petroleum and environmental geosciences. If we now define U as a uniform random variable $U = U(0,1)$, it follows that $X = F_x^{-1}(U)$. In other words, by equating the value of a uniform random variable between 0 and 1 to the CDF, we can back calculate the corresponding value of the random variable of interest. Fig. 6.9 demonstrates this concept for a single random variable, where the sampling process works as follows:

1. Generate n uniform random numbers u_1, u_2, u_3, ..., u_n from $U(0,1)$.
2. Solve for $x_i = F_x^{-1}(u_i)$, $i = 1, 2, ..., n$.

This results in a design containing n independent samples with replacement, i.e., the same value could end up being sampled multiple times. Variations on this approach could use different marginal distributions in the sampling of the inputs or possibly include draws from a joint distribution over subsets of inputs. See Tung and Yen (2005) for additional details on how this CDF-inverse method can be extended to the case of multiple random variables for both noncorrelated and correlated cases.

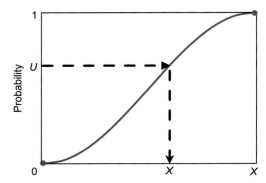

FIG. 6.9   Schematic of random sampling with the CDF inverse method.

Random designs are easy and straightforward to produce. However, they could also suffer from poor "space-filling" characteristics. That is, multiple observations can end up clustered in one part of the space and provide largely redundant information about the behavior of the response surface in that region. Other parts of the space may be sparsely populated, and the redundant observations could be put to better use filling in those gaps.

Latin Hypercube Sampling

A Latin hypercube sample (LHS) design is intended to fill the predictor space by randomly selecting observations in equal probability bins across the range of the inputs (McKay et al., 1979). It is a stratified sampling procedure that involves dividing the range for each input into strata of equal probability, picking one value from each interval, and randomly combining values picked for different variables. The sampling is done in such a way that for a sample of size n, there will be exactly one observation in each of the intervals $[0, 1/n]$, $[1/n, 2/n]$,..., $[(n-1)/n, 1]$ for each of the inputs. In practice, the $[0, 1]$ bounds on the values in LHS samples are interpreted to be a probability, and the design points are transformed through some probability distribution on the inputs. This has the effect of spreading the sampled points across equal regions of probability for each input (according to the chosen distribution), which also results in a reduction in the computed variance of the corresponding model outcome (Iman and Helton, 1985). Fig. 6.10 shows how the LHS scheme samples five different values from nonoverlapping probability bins for different variables and pairs them in a random manner.

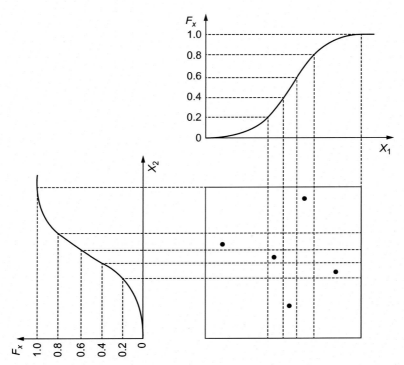

FIG. 6.10   **Latin Hypercube sampling in two variables with five samples.**

Correlation Control in LHS

Input-input correlations are important to take into account whenever (a) sensitive inputs are correlated, (b) the input-output model is nonlinear, and (c) tails of the output distribution are important. Correlations should be handled explicitly as functional relationships if the RCC (rank correlation coefficient) > 0.9. Otherwise, the restricted pairing technique (Iman and Conover, 1982), when used in conjunction with LHS, allows preserving any specified correlation structure between uncertain inputs and eliminating spurious input-input correlations. As noted earlier, the rank correlation is used for generating correlated samples as it is a more robust and distribution-free measure.

Let $\mathbf{T}$ be the actual correlation matrix, and $\mathbf{C}$ the desired one. We can define a transformation matrix $\mathbf{S}$ such that $\mathbf{STS}' = \mathbf{C}$. Here, $\mathbf{S}$ is any arbitrary matrix, and $\mathbf{S}'$ denotes its transpose. If we apply Cholesky factorizations (Press et al., 1992) to both $\mathbf{C}$ and $\mathbf{T}$, i.e., $\mathbf{C} = \mathbf{PP}'$, and $\mathbf{T} = \mathbf{QQ}'$, then it is easy to show that $\mathbf{S} = \mathbf{PQ}^{-1}$. Here, $\mathbf{P}$ is the lower triangular decomposition of the desired correlation matrix $\mathbf{C}$, and $\mathbf{Q}$ is the lower triangular decomposition for the actual correlation matrix $\mathbf{T}$. If $\mathbf{R}$ is the original matrix of ranks, then $\mathbf{R}^* = \mathbf{RS}'$ produces the desired correlation matrix $\mathbf{C}$. An illustrative example is presented below for a two-variable case (Fig. 6.11).

R			T			C	
3	2		1	−0.4		1	0.5
1	4		−0.4	1		0.5	1
5	1						
2	3		**Q**			**P**	
4	5		1	0		1	0
RCC	**−0.4**		−0.4	0.916515		0.5	0.866025

Q⁻¹			S			S'	
1	0		1	0		1	0.877964
0.436436	1.091089		0.877964	0.944911		0	0.944911

R* (raw)		R* (rank)			
3	4.523716	3	1		
1	4.657609	1	3		
5	5.334734	5	4	**RCC**	**0.5**
2	4.590662	2	2		
4	8.236414	4	5		

FIG. 6.11   Example showing Iman-Conover restricted pairing technique for inducing correlation with LHS.

6.3.2 Computational Considerations

Number of Samples

In the MCS process, it is important to ensure that multiple model computations are carried out using a sufficient number of sampled parameter vectors to obtain a stable solution for key performance indicators (e.g., mean, 90th percentile). One widely used rule of thumb for

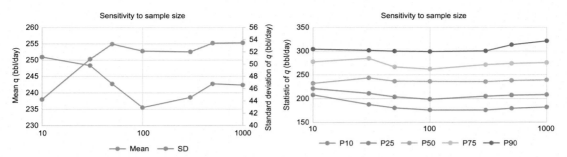

FIG. 6.12 Sensitivity to sample size in Monte Carlo simulation results for the exponential decline problem discussed in Example 6.2.

selecting an optimal sample size in LHS in order to obtain a stable mean is the 4/3N rule (Iman and Helton, 1985), where N is the number of uncertain inputs. However, additional runs may be required if tail percentiles are to be used as performance metrics of interest.

Fig. 6.12 shows the sensitivity to sample size for the Monte Carlo simulation version of the exponential decline problem discussed later in Example 6.2. The left panel shows the sensitivity of mean and standard deviation as the sample size is varied between 10 and 1000. The exaggerated scale notwithstanding, it is clear that stable values of both mean and standard deviation can be obtained with ~300 samples. The right panel shows a similar plot for the P10, P25, P50, P75, and P90 statistic values. Once again, the results appear stable around 300 samples, excepting for the P90 value. This lack of convergence for the extreme percentiles is a commonly observed situation, as noted earlier.

Visualization of Results

MCS results can be presented in a variety of ways. For time-independent output, it is generally recommended to show the CDF with two or three different sample sizes, as presented in Fig. 6.13. Here, the 300- and 1000-sample cases track each other very well, but

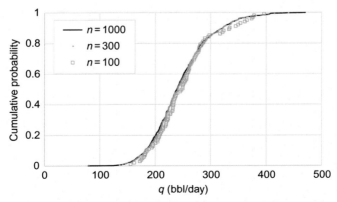

FIG. 6.13 Comparison of CDFs at three different sample sizes from Monte Carlo simulation results for the exponential decline problem discussed in Example 6.2.

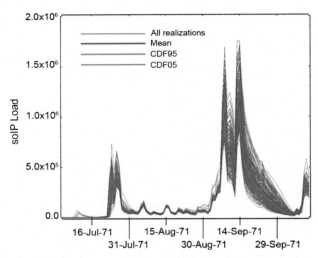

FIG. 6.14   **Graphic representation of Monte Carlo simulation results for time-dependent output.** *From Mishra, S., 2009. Uncertainty and sensitivity analysis techniques for hydrologic modeling. J. Hydroinf. 11 (3–4), 282–296.*

the 100-sample case shows clear divergence at both low and high percentiles (confirming the earlier results).

For time-dependent output, it is recommended that all of the probabilistic results should be presented in the form of a "horsetail" plot, along with the running mean and the 5th–95th percent confidence bounds. An example of such plots is shown in Fig. 6.14 for an environmental tracer transport problem (Mishra, 2009).

EXAMPLE 6.2 Monte Carlo simulation

Consider the problem of calculating the original oil in place for a new field and the associated uncertainty using the simple volumetric reserves estimation model:

$$N = \frac{7758Ah\phi(100 - S_{wi})}{10^4 B_{oi}} \tag{6.2}$$

where N is the original oil in place (stock tank barrels or STB), A is the total area for the reservoir (acres), h is average net pay thickness (ft), ϕ is average porosity (%), S_{wi} is initial water saturation, and B_{oi} is oil formation volume factor at the original pressure (reservoir barrels per stock tank barrels, or RB/STB). Using data from Murtha (1994) as given in Table 6.5 (INPUTS_TAB6-6.DAT), which represent the variability of the key input parameters over 26 fields in the Repetto basin, (a) fit appropriate distributions to the uncertain variables, (b) generate Monte Carlo simulation results for 100 and 500 Latin hypercube samples with and without correlation among the inputs, (c) test the validity of predicting P10, P50, and P90 for N using the equivalent percentiles for the five uncertain inputs.

TABLE 6.5 Input Data for Probabilistic Reserves Estimate Calculation

A (acres)	h (ft)	φ (%)	S_{wi} (%)	B_{oi} (RB/STB)
200	172	27	28	1.24
250	72	38	30	1.05
355	388	21	40	1.17
1268	125	32	35	1.04
388	224	20	37	1.3
265	250	20	37	1.3
445	332	26	25	1.16
525	338	29	27	1.16
144	95	36	40	1.08
365	133	32	25	1.04
1200	511	24	31	1.05
320	85	28	25	1.05
3000	250	36	36	1.05
445	150	38	35	1.05
1133	300	23	40	1.15
1133	400	32	23	1.1
1133	325	26	30	1.15
374	91	20	40	1.43
355	300	30	50	1.24
373	130	28	35	1.08
1000	80	33	19	1.05
859	123	33	19	1.05
270	80	34	18	1.05
400	50	35	18	1.05
200	75	30	26	1.05
180	325	25	37	1.01

Correlation Matrix With Raw Data

	A	h	φ	S_{wi}	B_{oi}
A	1				
h	0.29	1			
φ	0.18	−0.47	1		
S_{wi}	0.01	0.30	−0.38	1	
B_{oi}	−0.22	0.16	−0.68	0.46	1

Correlation Matrix With Ranks of Data

	A	h	φ	S_{wi}	B_{oi}
A	1				
h	0.33	1			
φ	0.04	−0.49	1		
S_{wi}	−0.14	0.35	−0.42	1	
B_{oi}	−0.11	0.42	−0.64	0.49	1

After Murtha, J.A., March 1, 1994. Incorporating historical data into Monte Carlo simulation. Soc. Pet. Eng. https://doi.org/10.2118/26245-PA.

Solution

Part (a)

The data were fitted with different parametric distribution models using the nonlinear regression procedure described previously in Section 6.2.2. The fitted models and parameters are the following:

A—lognormal	$\ln(A) = LN[\alpha = 6.11,\ \beta = 0.87]$
h—lognormal	$\ln(h) = LN[\alpha = 5.16,\ \beta = 0.84]$
ϕ—normal	$\phi = N[\mu = 29.57,\ \sigma = 6.58]$
S_{wi}—Weibull	$S_{wi} = W[\lambda = 35.21,\ k = 3.62]$
B_{oi}—Beta	$B_{oi} = B[\alpha = 1.78,\ \beta = 12.12] + 1$

The empirical CDFs for these parameters and their corresponding model fitted values are shown in Fig. 6.15, indicating good fits for all five variables.

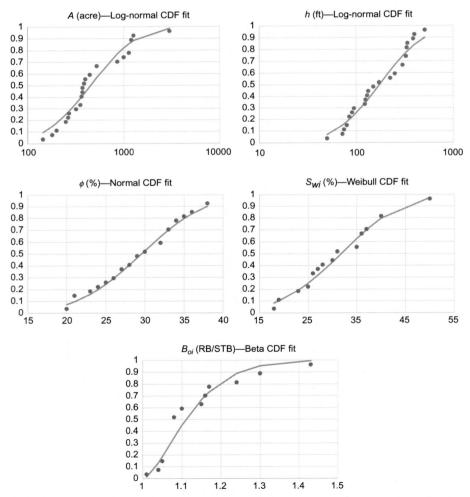

FIG. 6.15 **Comparison of empirical CDFs and corresponding model fits for uncertain inputs of interest, Monte Carlo simulation example. X-axis labels for each variable are given at the top; y-axis represents cumulative probability in all cases.**

Part (b)

Next, 500 Latin hypercube samples for the five variables were generated, taking into account the correlation structure of the dataset as measured by the rank correlation matrix (Table 6.5) (MCS_500_CORR.DAT). A scatterplot matrix for this case is presented in Fig. 6.16, which shows that the input-input correlation structure for the data was properly reproduced by the sampling algorithm.

Three other cases were generated: (a) 500 samples without correlation, (b) 100 samples with correlation, and (c) 100 samples without correlation. These realizations (i.e., vectors of sampled values for the five uncertain parameters) were used to the oil-in-place, N, using Eq. (6.2).

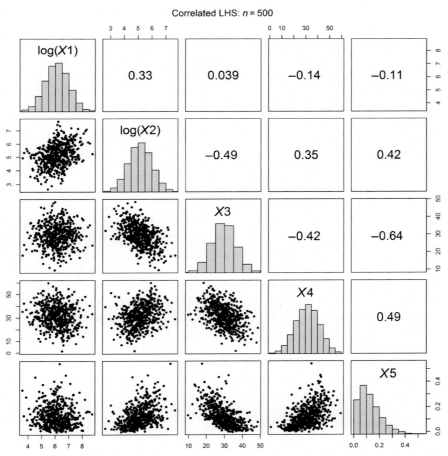

FIG. 6.16 Scatterplot matrix for the sampled data, 500 LHS samples with correlation. Here, $X1 = A$, $X2 = h$, $X3 = \phi$, $X4 = S_{wi}$, and $X5 = 1 - B_{oi}$.

TABLE 6.6   Key Statistics and Percentiles From Monte Carlo Simulation Results, 100 and 500 Samples, With and Without Correlation

	Rand 500	Rand 100	Corr 500	Corr 100
Mean	226.6	217.7	264.8	273.0
Median	107.3	101.2	113.9	101.9
St. dev	378.8	346.4	543.0	598.3
P5	11.4	15.0	11.5	8.4
P10	19.2	23.1	20.1	16.0
P50	107.3	101.2	113.9	101.9
P90	519.1	499.6	594.9	528.2
P95	858.7	701.4	917.6	852.4

Table 6.6 shows the key statistics and percentiles for all of the cases. The corresponding data files are (MCS_500_RAND.DAT), (MCS_100_CORR.DAT), and (MCS_100_RAND.DAT).

Fig. 6.17 (top) shows the Monte-Carlo-simulation-based CDFs for the 100 and 500 samples with correlation, suggesting that a sample size of 100 may be too small for this problem. Fig. 6.17 (bottom) compares the CDFs for the 500-sample case with and without correlation, clearly showing that the correlated case leads to more extreme results because of the combination of low-low and high-high values. This is also confirmed by the higher standard deviation and tail percentile values for the correlated versus uncorrelated cases given in Table 6.6.

Part (c)

Table 6.7 shows the P10, P50, and P90 values for the five uncertain inputs, based on the distributional parameters described earlier in Part (a). Also shown therein are the corresponding values for N calculated using Eq. (6.2) and the corresponding Monte Carlo simulation result using 500 uncorrelated samples. Clearly, the simple combination of percentiles produces a much wider distribution, with lower results compared with the Monte Carlo simulation at the sub-50 percentiles and higher results at the super-50 percentiles. Note, however, that the P50 result is reasonably reproduced—suggesting that this is the only case—where a combination of percentiles can be applied. The divergence between the two calculations is an artifact of implicitly assuming perfect correlation between the inputs, which is what the simplistic combination of percentiles essentially accomplishes, and is referred to as the "problem of compounded conservatism" in the risk analysis literature. For example, Bogen (1994) has shown that in a simple multiplicative model of risk, if upper p-fractile ($100p$th percentile) values are used for each of several statistically independent input variates, the resulting risk estimate will be the upper p'-fractile of risk predicted according to that multiplicative model, where $p' > p$. The difference between p' and p may be substantial, depending on the number of inputs, their relative uncertainties, and the value of p selected.

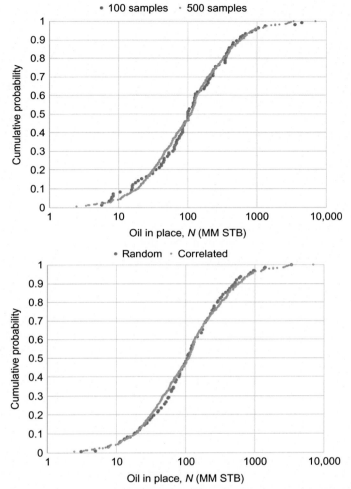

FIG. 6.17 **CDFs from Monte Carlo simulation showing effect of sample size (top) and input-input correlation (bottom).**

TABLE 6.7 Calculation Using Simple Combination of Percentiles

	A	h	f	S_{wi}	B_{oi}	N	N_MCS
P5	108.37	43.92	18.76	15.60	1.02	5.72	11.4
P10	148.23	59.47	21.17	18.99	1.03	11.35	19.2
P50	449.84	174.14	29.56	31.83	1.11	110.32	107.3
P90	1363.66	510.02	37.97	44.27	1.25	915.04	519.1
P95	1856.68	689.68	40.30	47.55	1.29	1621.99	858.7

6.4 UNCERTAINTY IMPORTANCE ASSESSMENT

In this section, we describe how to analyze the probabilistic calculations (i.e., uncertainty propagation results) to determine input-output relationships, and how to identify the key uncertain parameters. The motivation here is the fact that probabilistic models can consist of tens or hundreds of parameters that are uncertain, and whose interactions with one another are potentially complex and/or nonlinear. As such, it is difficult to develop a straightforward understanding of causal input-output relationships, critical uncertainties and key sensitivities based on a simple evaluation of model results. A systematic approach is therefore needed to extract this understanding.

6.4.1 Basic Concepts in Uncertainty Importance

The goal of *uncertainty importance assessment* (aka *global sensitivity analysis*), is to study how the variation in the probabilistic output of a model can be apportioned to different sources (Mishra et al., 2009). This is in contrast to traditional one-parameter-at-a-time *local sensitivity analysis* of subsurface flow and transport models, which involves perturbing each of the parameters by a small amount, one at a time, from a reference value and computing the corresponding change in the model output (Hill and Tiedeman, 2007). Sensitivity coefficients, computed as the change in output divided by the change in input, reflect the slope of the input-output relationship at the reference point. However, unless the functional relationship between the output and the input of interest is linear over the entire range of input values, such analyses can only provide information regarding the relative sensitivities of input parameters that is valid locally. The "one-off" nature of these analyses also precludes a proper accounting for synergistic effects between inputs. Global sensitivity analysis techniques have therefore emerged as an attractive alternative for investigating input-output sensitivities that are valid over the full range of parameter variations and parameter combinations considered in the analysis (Saltelli et al., 2000).

In the context of probabilistic modeling, global sensitivity analysis involves examining the relationship between uncertain model inputs and corresponding model outputs to answer such questions as the following: (a) which uncertain parameters or inputs have the greatest impact on the overall uncertainty (variance) in probabilistic model outcomes? and (b) what are the key factors controlling the separation of model results into extreme-outcome producing realizations? In addition to identification of key variables affecting uncertainty in predicted model outcomes, global sensitivity analyses results are also useful for verification of model performance (i.e., testing for physically reasonable results) and for providing feedback to data collection efforts for uncertainty/risk reduction (Mishra et al., 2009).

Uncertainty importance assessment is essentially a *variance-partitioning* problem. As shown schematically in Fig. 6.18, the contribution to output uncertainty (variance) by an input is a function of both the uncertainty of the input variable and the sensitivity of the output to that particular input. In general, input variables identified as important in global sensitivity analysis have both characteristics; they demonstrate significant variance and are characterized by large sensitivity coefficients. Conversely, variables that do not show up as important

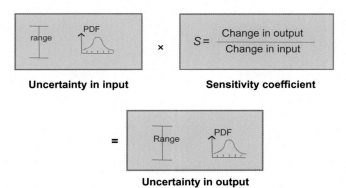

FIG. 6.18 **Uncertainty importance concepts.**

per these metrics either are restricted to a small range in the probabilistic analysis or are variables to which the model outcome does not have a high sensitivity.

Global sensitivity analysis on probabilistic modeling results can be viewed as a form of statistical data mining, i.e., the use of statistical techniques for extracting causal relationships, structures, patterns, and/or trends among dependent and independent variables in large-dimensional datasets (Hastie et al., 2008). Here, sampled values of uncertain inputs are treated as independent variables and computed model outputs are treated as dependent variables. In general, sampled inputs are considered to be time-independent. If the output of interest is time-dependent, then its values are extracted at fixed time slices for the analysis. As noted above, the objective of global sensitivity analysis is to develop input-output relationships and/or decision rules that capture the model behavior over the full range of inputs and corresponding model outcomes.

In this section, we will describe some basic uncertainty importance analysis techniques that are most appropriate in conjunction with sampling-based uncertainty analysis. These methods build upon the results of Monte-Carlo-simulation-based uncertainty analyses and hence, do not require resampling of the uncertain parameters and recomputation of model results. These are (1) scatterplots and rank correlation analysis and (2) stepwise rank regression and partial rank correlation analysis. We will also discuss other techniques such as entropy (mutual information) and classification tree analysis that can be useful for specialized situations.

6.4.2 Scatter Plots and Rank Correlation Analysis

Scatterplots were discussed in Section 2.2.3 as a means of graphically depicting bivariate relationships. In the context of Monte Carlo simulation, we can also use scatterplots for visual analysis of input-output relationships. They provide qualitative insights into the strength (e.g., strong or weak) and nature (e.g., linear or nonlinear) of cause-and-effect relationships. However, a systematic examination of scatterplots to identify the most influential input parameters is generally not feasible unless the model has only a few uncertain inputs.

Fig. 6.19 shows two of the input-output scatterplots for the oil-in-place uncertainty analysis problem discussed in the previous section. From the left panel, it is clear that the sampled

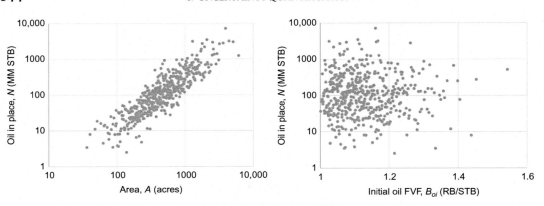

FIG. 6.19   **Input-output scatterplots from oil-in-place uncertainty analysis problem.**

value of A has a very strong influence on the computed value of N, as evidenced from the log-log plot. Conversely, the right panel shows that the sampled value of B_{oi} has minimal effect on the computed value of N.

It is more useful to use scatterplots in conjunction with rank correlation analysis to analyze uncertainty importance. Following the discussion in Chapter 2.4, we can express the Spearman's rank correlation coefficient, RCC, between any input-output pair as (Helton et al., 1991)

$$RCC[y, x_k] = \frac{\sum_k (x_k - \bar{x})(y_k - \bar{y})}{\left[\sum_k (x_k - \bar{x})^2 \sum_k (y_k - \bar{y})^2\right]^{1/2}} \tag{6.3}$$

where x is the rank-transformed input of interest, y is the rank-transformed output, the overbar symbol denotes the sample mean, and k is an index for the samples (realizations). Rank transformation, where variables are ranked in ascending order with their values replaced by the ranks, is the simplest nonparametric linearizing technique (Iman and Conover, 1983). The RCC provides a measure of the degree to which the input variable of interest and the output can change together. It quantifies the strength of linear and monotonic association between the input-output pair—with the rank transformation facilitating a linearization of any underlying nonlinear trends (Helton, 1993). Positive values of the RCC imply that an increase in the input corresponds to an increase in the output, with negative values implying the reverse situation. The larger the absolute value of the RCC, the stronger the relationship between the input-output pair, i.e., the greater the uncertainty importance. Fig. 6.20 shows a graphic comparison of the RCC values for the 500-sample uncorrelated case.

Note that the utility of the RCC is limited to the case of uncorrelated inputs. This point will be discussed further in the next section.

6.4.3 Stepwise Regression and Partial Rank Correlation Analysis

A popular framework for uncertainty importance analysis involves building a multivariate linear rank regression model of the form

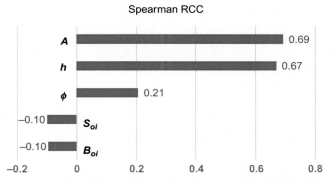

FIG. 6.20 **Input-output RCC bar charts from oil-in-place uncertainty analysis problem.**

$$\hat{y} = b_0 + \sum_j b_j x_j \tag{6.4}$$

where y (overbar) denotes the (predicted) rank-transformed output, the x_j are the rank-transformed input variables of interest, and the b_j are the unknown coefficients (Helton, 1993).

The regression coefficients are generally determined using a stepwise regression procedure (Draper and Smith, 1981). As discussed in Section 4.5, this involves the construction of a sequence of regression models starting with the parameter that explains the largest amount of variance in the output. At each successive step, the parameter that explains the largest fraction of residual variance is included in the model. The process continues until all input variables that explain statistically significant amounts of variance in the output have been included in the model.

When a linear additive input-output model such as Eq. (6.4) is built with uncorrelated inputs, the goodness-of-fit of the model can be expressed as (Draper and Smith, 1981)

$$R^2 = \sum_j RCC^2 [y, x_j] \tag{6.5}$$

where R^2, the coefficient of determination, denotes the fractional variance in y explained by the model. Thus, the term $RCC^2 [y, x_j]$ can be interpreted as the fractional variance in y explained by the jth independent variable. This is a more useful measure for interpreting the RCC in the context of input-output models to assign uncertainty importance.

When some of the input variables are correlated, the goodness-of-fit of the input-output model can no longer be expressed via a simple linear sum as in Eq. (6.5), but must also include terms reflecting the covariance of the correlated inputs. In such situations, it becomes difficult to assign a unique component of the output variance to each of the uncertain inputs. When variables are correlated, a more appropriate measure of uncertainty importance is the partial rank correlation coefficient (*PRCC*). *PRCCs* quantify the strength of a linear relationship between input-output pairs after eliminating the linear influence of other input variables (Draper and Smith, 1981).

The concept of partial rank correlation can be explained as follows. Let y denote the rank-transformed output variable and x_j, $j=1...n$, denote the rank-transformed uncertain inputs—some of which may be correlated. In order to determine the *PRCC* between y and the p-th uncertain input, x_p, a linear regression model is first built between y and all the other uncertain inputs, with y_{p_fit} denoting the regression-fitted variable. Next, a linear regression model is built between x_p and all the other uncertain inputs, with x_{p_fit} denoting the regression-fitted variable. The RCC between the residuals arising out of these regressions will be free from the effects of input-input correlations, and is defined as the PRCC (Draper and Smith, 1981):

$$PRCC[y, x_p] = RCC[y - y_{p\_fit}, x - x_{p\_fit}] \tag{6.6}$$

RamaRao et al. (1998) showed that the square of the PRCC can be interpreted as the gain in R^2 of an input-output regression model—when the selected variable is brought into regression—as a fraction of the currently unexplained variance. PRCCs can be readily calculated from the input-input correlation matrix and the input-output correlation vector using simple matrix algebra. This practical strategy, which does not involve building a sequence of regression models, starts with the augmented correlation matrix between the output variable, y, and the independent variables x_j, $j = 1,...., n$, written as

$$C = \begin{bmatrix} 1 & r_{12} & \cdots & r_{1n} & r_{1y} \\ r_{21} & 1 & \cdots & r_{2n} & r_{2y} \\ \cdots & \cdots & \cdots & \cdots & \cdots \\ r_{n1} & r_{n2} & \cdots & 1 & r_{ny} \\ r_{y1} & r_{y2} & \cdots & r_{yn} & 1 \end{bmatrix} = \begin{bmatrix} A & B \\ B^T & 1 \end{bmatrix} \tag{6.7}$$

where the matrix A represents the input-input correlation matrix with elements $r_{ij} = RCC[x_i, x_j]$ and the vector B^T denotes the output-input correlation vector with elements $r_{yj} = RCC[y, x_j]$. As shown by RamaRao et al. (1998), the PRCC between x_j and y can be obtained from the elements of C^{-1}, the inverse of C, as

$$PRCC[y, x_j] = -\frac{c_{jy}}{\sqrt{c_{jj}c_{yy}}} \tag{6.8}$$

where the subscript y is now used as the designator for row and column $n+1$ in C^{-1}.

EXAMPLE 6.3 Uncertainty importance analysis

The Monte Carlo simulation results for the 100-sample uncorrelated case and the 500-sample correlated case from Example 6.2 are given in (MCS_100_RAND.DAT) and (MCS_500_CORR.DAT). Determine the relative importance of various uncertain inputs using both RCC and PRCC.

Solution

For the 500-sample correlated case, the augmented correlation matrix is given by

	X1	X2	X3	X4	X5	Y
X1	1	0.330025	0.039105	−0.13718	−0.10916	0.878514
X2	0.330025	1	−0.48913	0.348095	0.419246	0.663484
X3	0.039105	−0.48913	1	−0.4179	−0.63754	−0.03724
X4	−0.13718	0.348095	−0.4179	1	0.487029	−0.07616
X5	−0.10916	0.419246	−0.63754	0.487029	1	−0.02485
Y	0.878514	0.663484	−0.03724	−0.07616	−0.02485	1

The inverse of this matrix is given by

14.34689	10.49055	2.599638	−1.68031	−0.84375	−19.6164
10.49055	11.24454	2.888181	−1.96581	−1.18653	−16.7483
2.599638	2.888181	2.484343	−0.25393	0.678332	−4.11006
−1.68031	−1.96581	−0.25393	1.704022	−0.27915	2.893845
−0.84375	−1.18653	0.678332	−0.27915	2.013093	1.582517
−19.6164	−16.7483	−4.11006	2.893845	1.582517	29.45223

The $PRCC$s can now be calculated using Eq. (6.8). For example, the $PRCC$ between Y and $X1$ is given by $PRCC_{Y-X1} = -(-19.6164)/\text{sqrt}(14.34689*29.45223) = 0.954292$. The $PRCC$s for the other output-input pairs and the corresponding RCC values are given below:

	Y–X1	Y–X2	Y–X3	Y–X4	Y–X5
PRCC	0.954	0.920	0.480	−0.408	−0.206
RCC	0.879	0.663	−0.037	−0.076	−0.025

The importance of variables $X1$, $X2$, and $X5$ is the same with both methods, and that of variables $X3$ and $X4$ has been switched around. However, given the finite correlation present in many elements of the input-input correlation matrix, the importance ranking from $PRCC$s should be considered more reliable.

For the 100-sample uncorrelated case, the augmented correlation matrix is given by

	X1	X2	X3	X4	X5	Y
X1	1	−0.01261	0.044896	−0.00469	0.034443	0.711347
X2	−0.01261	1	−0.02523	0.056958	0.05679	0.624974
X3	0.044896	−0.02523	1	0.165833	−0.26056	0.203732
X4	−0.00469	0.056958	0.165833	1	−0.15777	−0.0461
X5	0.034443	0.05679	−0.26056	−0.15777	1	−0.01041
Y	0.711347	0.624974	0.203732	−0.0461	−0.01041	1

From the inverse of this matrix, the $PRCC$s can be calculated using Eq. (6.8) as before for all input-output pairs. These values and their corresponding RCC values are given below:

	Y − X1	Y − X2	Y − X3	Y − X4	Y − X5
PRCC	0.960	0.952	0.674	−0.488	−0.177
RCC	0.711	0.625	0.204	−0.046	−0.010

As expected, the importance ranking (based on the absolute value of either the $PRCC$ or RCC) is the same in both cases.

It should be pointed out that the actual values of the $PRCC$s are not as easy to interpret as the RCCs, which are related to the slope of the best-fit line through a rank-transformed input-output scatterplot. While the relative magnitude of the $PRCC$s are important indicators of variable importance, the numeric values only have a specific meaning in the context of building a multivariate input-output regression model. As noted earlier, the square of the $PRCC$ gives the increase in R^2, when a new variable is added, as a fraction of the currently unexplained variance in the model. From a practical standpoint, ranking the variables with *PRCCs* and examining scatterplots to understand input-output relationships would be a reasonable strategy for sensitivity analysis of probabilistic models when inputs are correlated.

A related measure of variable importance is based on the concept of R^2-loss, i.e., the loss in explanatory power of an input-output model if a particular variable is excluded. The larger the R^2-loss, the greater the importance of the variable of interest. Sections 8.3.3 and 8.4.3 provide additional details on how this approach can be applied in practice.

6.4.4 Other Measures of Variable Importance

Entropy (Mutual Information) Analysis

Since the concept of correlation and regression is strictly applicable to monotonic relationships, it is useful to pose the uncertainty importance problem in the general terms of identifying important nonrandom patterns of association. An example of such a situation would be if the performance metric of interest is quadratic in nature (as is likely to be the case in the context of history matching). Here, the word "association" is used in a broader context than "correlation" and includes both monotonic and nonmonotonic relationships. Determining the significance and strength of input-output association is facilitated by the information-theoretic concept of entropy, which provides a useful framework for the characterization of uncertainty (or information) in the univariate case, and redundancy (or mutual information) in the multivariate case (e.g., Press et al., 1992). Mishra and Knowlton (2003) describes a methodology for global sensitivity analysis that combines the mutual information concept with contingency table analysis.

As per Press et al. (1992), let the input variable x have I possible states (labeled by i), and the output variable y have J possible states (labeled by j). This information can be compactly organized in terms of a contingency table, a table whose rows are labeled by the values of the independent variable, x, and whose columns are labeled by the values of the dependent variable, y. The entries of the contingency table are nonnegative integers giving the number of observed outcomes for each combination of row and column. The corresponding probabilities are readily obtained by normalization.

The mutual information between x and y, which measures the reduction in uncertainty of y due to knowledge of x (or vice versa), is defined as (e.g., Bonnlander and Weigend, 1994)

$$I(x,y) = \sum_i \sum_j p_{ij} \ln \frac{p_{ij}}{p_{i\cdot} p_{\cdot j}} \tag{6.9}$$

Here, p_{ij} is the probability of outcomes corresponding to both state x_i and state y_j, while $p_{i\cdot}$ is the probability of outcomes corresponding to state x_i alone, and $p_{\cdot j}$ is the probability of outcomes corresponding to state y_j alone. A useful measure of importance defined on the basis of mutual information is the so-called R-statistic (Granger and Lin, 1994):

$$R(x,y) = [1 - \exp\{-2I(x,y)\}]^{1/2} \tag{6.10}$$

R takes values in the range $[0,1]$, with values increasing with I. R is zero if x and y are independent and is unity if there is an exact linear or nonlinear relationship between x and y.

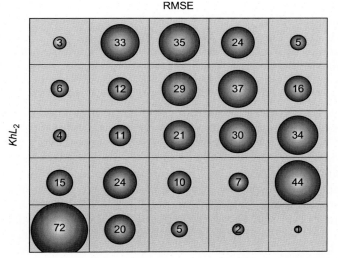

FIG. 6.21 **Bubble plot showing visualization of contingency table for a key nonmonotonic input identified from entropy analysis.** *From Mishra, S., Deeds, N., Ruskauff, G., 2009. Global sensitivity analysis techniques for probabilistic ground water modeling. Ground Water 47, 727–744. https://doi.org/10.1111/j.1745-6584.2009.00604.x.*

In summary, the entropy-based measure *R*-statistic is a very general tool for quantifying the strength of an association, even in nonlinear and/or nonmonotonic cases. Mishra et al. (2009) show how important nonmonotonic patterns, missed by stepwise regression analysis, can be readily identified using entropy analysis (Fig. 6.21). Here, the performance metric of interest is the RMSE for a groundwater model calibration problem. The bubble plot corresponding to the contingency table clearly reveals a significant association, as indicated by the inverted V-shaped pattern and also quantified by a *R*-statistic of 0.691. However, the corresponding *RCC* is only 0.09, reflecting the inability of linear correlation to capture the strength of a nonmonotonic relationship.

Classification Tree Analysis

Uncertainty importance analyses based on stepwise regression or mutual information concepts are typically applied to the entire spectrum of input-output data. However, specialized approaches may be required for examining small subsets (e.g., top and bottom deciles) of the output. To this end, classification tree analysis can provide useful insights into what variable or variables are most important in determining whether outputs fall in one or the other (extreme) category (Breiman et al., 1984). Such categorical problems may arise in the context of model calibration, where the factors contributing to good v/s poor fits may be of interest. Another example is contaminant transport, where insights on variables responsible for high v/s low migration distance may be useful. Classification tree analysis is also an important tool for data-driven modeling as discussed in Chapter 8.

A binary decision tree is at the heart of classification tree analysis. The decision tree is generated by recursively finding the variable splits that best separate the output into groups where a single category dominates. For each successive fork of the binary decision tree, the algorithm searches through the variables one by one to find the optimal split within each variable. The splits are then compared among all the variables to find the best split for that fork. The process is repeated until all groups contain a single category. In general, the variables that are chosen by the algorithm for the first several splits are most important, with less important variables involved in the splitting near the terminal end of the tree.

A common tree-building methodology is based on a probability model approach (Venables and Ripley, 1997). The classification tree is built by successively taking the maximum reduction in deviance over all the allowed splits of the leaves to determine the next split. The deviance is simply a measure of mean square error (for continuous response) or negative log-likelihood (for discrete response). Termination occurs when the number of cases at a node drops below a set minimum or when the maximum possible reduction in deviance for splitting a particular node drops below a set minimum. See Hastie et al. (2008) for additional details regarding the tree-building process and measures of importance.

In summary, classification tree analysis is a powerful tool for determining variable importance for categorical problems. Compared with linear regression modeling, tree-based models are attractive because (a) they are adept at capturing nonadditive behavior, (b) they can handle more general interactions between predictor variables, and (c) they are invariant to monotonic transformations of the input variables. Mishra et al. (2003) describe an application of this methodology for identifying key variables affecting extreme outcomes in a groundwater-driven radionuclide transport model. Fig. 6.22 shows a decision tree from a probabilistic analysis of groundwater model calibration, where the top 10% and bottom 10% of the realizations in terms of RMSE of the model fit are analyzed to identify the key variables affecting the spread in RMSE.

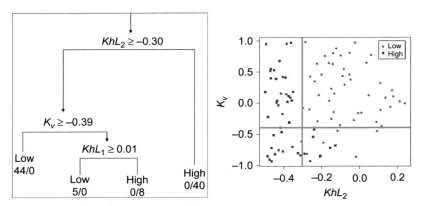

FIG. 6.22   **Classification tree example, showing decision tree (left), and partition plot (right).** *From Mishra, S., Deeds, N., Ruskauff, G., 2009. Global sensitivity analysis techniques for probabilistic ground water modeling. Ground Water 47, 727–744. https://doi.org/10.1111/j.1745-6584.2009.00604.x.*

6.5 MOVING BEYOND MONTE CARLO SIMULATION

As noted earlier, the main disadvantage with the MCS technique is the need to perform multiple model calculations. For large and/or complex models, the computational burden for a Monte Carlo analysis can be prohibitive. Engineers therefore commonly resort to using only a limited number of realizations as a computational shortcut, even though there is no assurance that the final results will be statistically reliable. An alternative approach is to use simplified or surrogate models using experimental design and response surface as discussed in Chapter 7 (Carreras et al., 2006). A second disadvantage concerns the issue of data availability for defining the range and distributions of the uncertain inputs. In many real-life situations, the lack of data often forces the engineer to make simplifying assumptions about the ranges and shape of the input distributions. Under such circumstances, the justification for using a full-blown MCS study, based on subjective assumptions about data distributions, becomes questionable at best. Finally, MCS may not be the most efficient strategy when the probability associated with only a limited number of model outcomes is desired.

In the following sections, we will discuss several alternatives to Monte Carlo simulation that address some of these shortcomings. These include the first-order second-moment method (FOSM), the point-estimate method (PEM), and the logic tree analysis (LTA).

6.5.1 First-Order Second-Moment Method (FOSM)

Often, uncertainty about model inputs is available only in terms of the first few statistical moments (e.g., mean and variance). Given this limited information, it is useful to ask if the corresponding uncertainty in model predictions can also be quantified in terms of the mean and the variance rather than the full distribution. The first-order second-moment method (FOSM) is one such methodology (Morgan and Henrion, 1990; Tung and Yen, 2005). As we shall see later, the FOSM approach is also the basis for the widely used error propagation formulas in experimental work.

General Expressions for Mean and Variance

Consider an uncertain quantity, F, which depends on the parameter vector, $\mathbf{x} = (x_1, x_2, .., x_i, .., x_N)$. A first-order Taylor expansion around the mean point, $\hat{\mathbf{x}}$, gives

$$F(\mathbf{x}) \cong F(\hat{\mathbf{x}}) + \sum_i \frac{\partial F}{\partial x_i}\bigg|_{\hat{\mathbf{x}}} (x_i - \hat{x}_i) \tag{6.11}$$

with $\hat{\mathbf{x}}$ being the vector of mean values of the uncertain parameters, where the partial derivatives in Eq. (6.11) are also evaluated. Taking the expected value of both sides of this expression yields

$$E[F] \cong F(\hat{\mathbf{x}}) + \sum_i \frac{\partial F}{\partial x_i}\bigg|_{\hat{\mathbf{x}}} E[x_i - \hat{x}_i] \tag{6.12}$$

where $E[\bullet]$ denotes the expectation (averaging) operator. Now, assuming small and uniformly distributed random parameter perturbations around the mean values such that the expectation term in the RHS can be dropped and all higher-order terms neglected, we obtain

$$E[F] \cong F(\hat{\mathbf{x}}) = F(\hat{x}_1, \hat{x}_2, .., \hat{x}_i, .., \hat{x}_N) \qquad (6.13)$$

Thus, the first-order estimate of the expected value (mean) of the uncertain quantity, F, is obtained simply by using the mean (expected value) of each of the uncertain parameters.

The variance of F is defined as

$$V[F] = \sigma_F^2 = E\left[(F - E[F])^2\right] \qquad (6.14)$$

which can be calculated by substituting Eqs. (6.10) and (6.12) in Eq. (6.13) as follows:

$$
\begin{aligned}
V[F] &\cong E\left[\left\{\left(F(\hat{\mathbf{x}}) + \sum_i \left.\frac{\partial F}{\partial x_j}\right|_{\hat{\mathbf{x}}} (x_i - \hat{x}_i)\right) - F(\hat{\mathbf{x}})\right\}^2\right] \\
&\cong \sum_i \sum_j \cong \sum_i \sum_j \left.\frac{\partial F}{\partial x_i}\frac{\partial F}{\partial x_j}\right|_{\hat{\mathbf{x}}} Cov\left[x_i x_j\right]
\end{aligned}
\qquad (6.15)
$$

where the covariance, $Cov[x_i x_j] = \rho[x_i x_j]\sigma[x_i]\sigma[x_j]$ can also be expressed in terms of the parameter correlation coefficients, ρ_{ij}, and the individual parameter standard deviations, σ. The variance of F is thus seen to depend on the variance-covariance relation of the input parameters and its sensitivity to the uncertain inputs.

For uncorrelated parameters, that is, when $\rho(x_i x_j) = 0$, the expression for variance simplifies to

$$V[F] \cong \sum_i \left(\left.\frac{\partial F}{\partial x_i}\right|_{\hat{\mathbf{x}}}\right)^2 V[x_i] \qquad (6.16)$$

because $Cov[x_i x_i] = V[x_i]$. Each term in the summation of Eq. (6.16) can be interpreted as the fractional contribution of x_i to the total variance of F. It may be recognized that Eq. (6.16) is also the commonly used expression for propagating experimental errors (Morgan and Henrion, 1990), and is conceptually equivalent to Fig. 6.18.

The sensitivity coefficients (partial derivatives) needed in Eqs. (6.15) and (6.16) for evaluating the variance of F can be computed either analytically or numerically. For simple petroleum geoscience problems, closed-form expressions for the derivatives can be readily generated. For complex models such as reservoir simulators, a forward difference calculation of the derivatives is often the common practical solution. In such cases, $(n + 1)$ functional evaluations are required for FOSM estimates of the mean and variance where n is the number of uncertain inputs. Thus, the FOSM technique can be competitive with Monte Carlo simulation so long as $n \sim 10$, rather than $n \sim 100$.

The first-order estimate of the mean given in Eq. (6.13) is a reasonable approximation so long as parameter variances are small and the function is only mildly nonlinear, which allows higher-order terms to be dropped. If these conditions are not met, then second-order terms need to be retained in the Taylor expansion of Eq. (6.11), leading to a correction to the mean that depends on the parameter covariance and mixed second partial derivatives (Dettinger and Wilson, 1981). Second-order corrections to the variance, which involve complex mixed partial derivatives of higher order, are rarely applied in practice (Morgan and Henrion,

1990). It is also advisable to perform variable transformations as needed so that the input-output relationship is (quasi) linear and parameter uncertainties are approximately symmetrical around the mean values (Tung and Yen, 2005).

EXAMPLE 6.4 FOSM application for exponential decline problem

Consider the exponential decline problem, $q = q_o \exp(-at)$, where q_o is initial oil rate (bbl/d), a is decline rate (1/year), and t is time (year). Given $E[q_o] = 650$ bbl/d, $\sigma[q_o] = 50$ bbl/d, $E[a] = 0.1$ 1/year, $\sigma[a] = 0.02$, and $\rho[q_o a] = -0.5$, estimate $E[q]$ and $\sigma[q]$ when $t = 10$ years.

Solution

First, we evaluate the expected value of q using Eq. (6.13):

$$E[q] = E[q_o] \exp\{-E[a] \cdot t\}$$

$$= (650) \exp\{-(0.10)(10)\}$$

$$= 239 \text{bbl/d}$$

Next, we develop analytic expressions for the required partial derivatives:

$$\partial q / \partial q_o = \exp(-at) = q/q_o$$

$$\partial q / \partial a = -q_o t \exp(-at) = -qt$$

Now, application of Eq. (6.15), with all quantities evaluated at the mean point, gives

$$V[q] = (\partial q / \partial q_o)^2 V[q_o] + (\partial q / \partial a)^2 V[a] + 2(\partial q / \partial q_o)(\partial q / \partial a) Cov[q_o a]$$

$$= (q/q_o)^2 V[q_o] + (qt)^2 V[a] + 2(q/q_o)(-qt)\rho[q_o a]\sigma[q_o]\sigma[a]$$

$$= (239/650)^2 (50)^2 + (239 \times 10)^2 (0.02)^2 + 2(239/650)(239 \times 10)(0.5)(50)(0.02)$$

$$= 338 + 2284 + 879 = 3501$$

$$\sigma[q] = \sqrt{3501} = 59.2 \text{bbl/d}$$

Error Analysis in Additive and Multiplicative Models

Additive models are those that take the general form

$$F = ax + by + cz \tag{6.17}$$

where x, y, and z are uncertain parameters and the coefficients a, b, and c are constant. For simplicity and without any loss of generality, Eq. (6.17) has been restricted to only three independent variables. Now applying Eq. (6.13), we get an expression for the mean

$$E[F] = aE[x] + bE[y] + cE[z] \tag{6.18}$$

Using Eq. (6.15), the variance of F is obtained as

$$V[F] = a^2 V[x] + b^2 V[y] + c^2 V[z] + 2ab Cov[xy] + 2bc Cov[yz] + 2ac Cov[zx] \tag{6.19}$$

with the last three terms dropping out if the uncertain parameters are uncorrelated. It is worth noting that both Eqs. (6.18) and (6.19) are exact because of the linearity of Eq. (6.17), which in turn makes the first-order Taylor expansion exact. Another important observation is that variances are additive (albeit as a weighted sum), but not standard deviations (i.e., error estimates).

The central limit theorem states that the sum of independent random variables will be normally distributed. Thus, knowing the mean and variance of the additive model given in Eq. (6.15) and assuming normality for F, we can estimate any quantile as shown in Example 3.6.

Multiplicative models are those that take the general form

$$F = \left[(x^a)(y^b)(z^c) \right] \tag{6.20}$$

where x, y, and z are uncertain parameters and the exponents a, b, and c are constant. As in the case of the additive model, Eq. (6.20) has been restricted to only three terms for simplicity and without any loss of generality. We can rewrite this equation as

$$\ln(F) = a\,\ln(x) + b\,\ln(y) + c\,\ln(z) \tag{6.21}$$

using a logarithmic transformation that converts our nonlinear multiplicative model into a linear additive one. Expressions for the mean and variance of $\ln(F)$ can now be readily derived using Eqs. (6.18) and (6.19), provided the moments of $\ln(x)$ are available. This is a useful approach to take when x, y, and z can be described using lognormal distributions.

As an alternative, an expression for the mean can also be derived using Eq. (6.13)

$$E[F] \cong \left[(E[x])^a\,(E[y])^b\,(E[z])^c \right] \tag{6.22}$$

assuming that the variables are independent. In order to calculate the variance using Eq. (6.15), we evaluate the partial derivatives at the mean point

$$\frac{\partial F}{\partial x} = \frac{a}{E[x]}E[F]; \quad \frac{\partial F}{\partial y} = \frac{b}{E[y]}E[F]; \quad \frac{\partial F}{\partial z} = \frac{c}{E[z]}E[F] \tag{6.23}$$

This leads to

$$V[F] \cong \left(\frac{a}{E[x]}E[F] \right)^2 V[x] + \left(\frac{b}{E[y]}E[F] \right)^2 V[y] + \left(\frac{c}{E[z]}E[F] \right)^2 V[z] \tag{6.24}$$

Dividing both sides by $E^2[F]$, and recalling that the coefficient of variation, $CV[x] = \sigma[x]/E[x]$, we can rewrite Eq. (6.24) as

$$CV^2[F] \cong a^2\,CV^2[x] + b^2\,CV^2[y] + c^2\,CV^2[z] \tag{6.25}$$

This is a very useful expression for estimating the relative error (i.e., coefficient of variation) in the output of a model as a function of the relative error of the inputs. Note that the square of the relative errors is additive (albeit as a weighted sum), but not the relative errors themselves.

The product of independent random variables can be shown to follow a lognormal distribution as per the central limit theorem. If the mean and variance of $(\ln F)$ are known, then any quantile of F can be estimated using the lognormal distribution relations as shown in Example 3.7.

EXAMPLE 6.5 Error analysis with multiplicative model

Consider the problem of estimating permeability from the slope of a Horner plot via the equation: $kh = 162.6 \, q\mu B/m$, where k is permeability (mD), h is thickness (ft), q is oil rate (bbl/d), μ is viscosity (cp), B is formation volume factor (rb/bbl), and m is Horner slope (psi/log-cycle). If the relative errors (coefficient of variation) in q, μ, and B are 10%, and the relative error in m is 20%, what is the relative error in the estimated value of the permeability-thickness product?

Solution

Our basic model is a multiplicative one similar to Eq. (6.20), with exponents +1 or −1. This simplifies the application of Eq. (6.17) to

$$CV^2[kh] = CV^2[q] + CV^2[\mu] + CV^2[B] + CV^2[m]$$
$$= (0.1)^2 + (0.1)^2 + (0.1)^2 + (0.2)^2 = 0.01 + 0.01 + 0.01 + 0.04 = 0.07$$
$$CV[kh] = \sqrt{0.07} = 0.26 = 26\%$$

Note that the actual values of the various parameters are not required for this error analysis— only the magnitude of the relative errors (i.e., standard deviation normalized by the mean). Also, note that if the errors are expressed in the $\{E[x] \pm \sigma[x]\}$ format, they need to be converted into coefficient of variations for Eq. (6.25) to be applied.

In summary, the FOSM technique is an appealing alternative to MCS when only the mean and variance of model outputs are of interest rather than the full CDF. It involves considerably less computational effort for problems with a small number of uncertain parameters, while providing results of comparable accuracy for linear and mildly nonlinear problems (Mishra and Parker, 1989; James and Oldenburg, 1997; Hirasaki, 1975).

6.5.2 Point Estimate Method (PEM)

Although the FOSM technique is conceptually simple, it has limited practical applicability for nonlinear models or models where numerical computation of derivatives could be burdensome. To overcome these limitations and to provide an efficient method for relating the statistical moments of the inputs to the moments of the output, the point-estimate method (PEM) was proposed by Harr (1989). In this method, the model is evaluated at a discrete set of points in the uncertain parameter space, with the mean and variance of model predictions computed using weighted averages of these functional evaluations.

The starting point in PEM is the estimation of the eigenvalues (λ_i) and eigenvectors (e_{ij}) of the correlation matrix for the uncertain variables. Each variable, x_j, is then perturbed around its mean by a factor, Δx_j:

$$\Delta x_j = \pm e_{ij} \sqrt{N} \sigma[x_j] \tag{6.26}$$

where N is the number of uncertain variables and σ denotes the standard deviation. The method thus results in $2N$ point estimates of the model, based on which the output mean is computed as follows:

$$E[F] = \sum_i \left(F_i^+ + F_i^- \right) \frac{\lambda_i}{2N} \qquad (6.27)$$

and the output variance is computed from

$$E\left[F^2\right] = \sum_i \left[\left(F_i^+ \right)^2 + \left(F_i^- \right)^2 \right] \frac{\lambda_i}{2N} \qquad (6.28)$$

by noting the relationship: $V[F] = E[F^2] - (E[F])^2$. Here F_i^+ and F_i^- denote estimates of model output corresponding to the perturbation of each input parameter from its mean value by Δx_j in the positive and negative directions and λ_i are the eigenvalues corresponding to each input (Fig. 6.23).

Although $2N$ model evaluations are required to compute the mean and variance as per Eqs. (6.27) and (6.28), it has been noted that in many cases the eigen transformation of the correlation matrix results in only a few dominant eigenvalues (Harr, 1989). Thus, it is possible to use this subset of eigenvalues for uncertainty propagation without any significant loss of accuracy.

In summary, the PEM approach is a derivative-free alternative to FOSM for estimating the mean and variance of uncertain model outputs. The original PEM algorithm of Harr (1989) was designed for correlated random variables with normal distributions. Chang et al. (1997) describe a methodology for extending this method to problems involving multivariate nonnormal random variables. Unlu et al. (1995) and Mishra (1998, 2000) provide comparative assessments of FOSM and PEM for uncertainty propagation using subsurface flow and transport models.

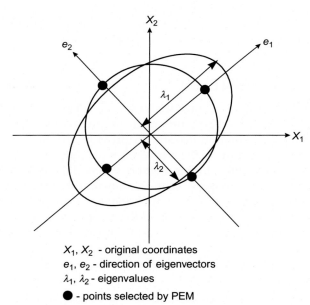

X_1, X_2 - original coordinates
e_1, e_2 - direction of eigenvectors
λ_1, λ_2 - eigenvalues
● - points selected by PEM

FIG. 6.23   **Selection of points for model evaluation in PEM.**

6.5.3 Logic Tree Analysis (LTA)

Logic tree analysis (LTA) is particularly useful for uncertainty propagation when parameter uncertainty is described using a limited number of probable states (e.g., high, medium, and low values), and their likelihoods. Logic trees (also known as probability trees) combine individual scenarios resulting from uncertain discrete events and/or parameter states. As such, they may be recognized as a special case of decision trees containing only chance nodes but no decision nodes (Morgan and Henrion, 1990).

The logic tree is organized such that independent effects are placed to the upstream (left) side, and dependent effects are placed to the downstream (right) side. Each branch is assigned a probability that is conditional on the values of the previous branches leading to that node. All scenarios must be considered in building the tree, so that probabilities for branches originating from each node sum to 1.

Consider a simple groundwater contaminant transport modeling problem involving two uncertain inputs: source concentration (s) and groundwater velocity (v). Uncertainty in the source node is represented by two values, $s1$ and $s2$, with probabilities $P1$ and $P2$, respectively. Uncertainty in the velocity node is also represented by two values, $v1$ and $v2$. These values have conditional probabilities ranging from $P3$ to $P6$, depending on which branch of the source node they are attached to. Each path from the root to an end branch (or terminal node) of the tree represents a feasible scenario. The four feasible scenarios for this system can be enumerated as ($s1,v1$), ($s1,v2$), ($s2,v1$), and ($s2,v2$). The probability of each scenario is the product of conditional probabilities of the branches along that path, as shown in Fig. 6.24.

The logic tree thus organizes various parameter combinations and their probabilities. Given this information, the computation of the consequence for each of the discrete combinations is a straightforward task. The results can be organized in terms of a table or graph of sorted discrete outcomes versus the corresponding summed probabilities. Such a "risk profile" is equivalent to a cumulative distribution of model output generated via MCS.

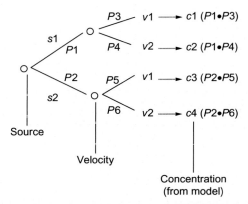

FIG. 6.24 **Schematic of logic tree construction and assessment of probabilities.**

In summary, the LTA methodology is a useful alternative to MCS when uncertainty characterization is based on a limited number of possibilities (as in the case of expert elicitation). Given the combinatorial nature of the algorithm, it can only handle a limited number of uncertain inputs, and is often useful in screening-type analyses. An example application of the LTA methodology is in risk assessments of the potential nuclear waste repository at Yucca Mountain, NV (Kessler and McGuire, 1999).

EXAMPLE 6.6 PEM and LTA applications for the exponential eecline problem

For the exponential decline problem discussed earlier in Example 6.1, calculate (a) mean, variance, and CDF (assuming normal distribution) using PEM and (b) CDF, mean, and variance using LTA. Assume that there is no correlation between the parameters. Compare these two CDFs with MCS simulation results from Mishra (1998).

Solution

(a) Point Estimate Method

The application of PEM was simplified for this problem because the parameters were taken to be uncorrelated. Thus, using eigenvalues of $\lambda_1 = 1.0$ and $\lambda_2 = 1.0$, with eigenvectors $(1,0)^T$ and $(0,1)^T$, we first calculate the perturbation for each variable, Δx_j (using Eq. 6.26), the resulting evaluation point $x_j(+)$ and $x_j(-)$, and the corresponding value of the function $F_i(+)$ and $F_i(-)$ as shown below:

λ_i		Δx_j	x_j (+)	x_j (−)	F_i (+)	F_i (−)	F_i^2 (+)	F_i^2 (−)
	Δq_o	70.71068	7.21E + 02	5.79E + 02				
1	Δa	0.0000	0.1000	0.1000	265.1346	213.1086	70296.38	45415.29
	Δq_o	0	6.50E + 02	6.50E + 02				
1	Δa	0.0283	0.1283	0.0717	180.2112	317.2896	32476.09	100672.7

Then, applying Eqs. (6.27) and (6.28), we can calculate the mean and standard deviation as

$E[q] = 243.9\,\text{bbl/d}$
$\sigma[q] = 52.1\,\text{bbl/d}.$

Recall that the corresponding FOSM values are $E[q] = 239.1$ bbl/d and $\sigma[q] = 52.1$ bbl/d from Example 6.2, after modifying the results for zero input-input correlation.

(b) Logic Tree Analysis

The first task is to approximate the continuous distributions assumed for q_o and a into discrete states. Following Clemen (1997), we note that for a symmetric distribution, an equivalent three-point distribution that preserves the first two statistical moments corresponds to the following:

$P = 0.185\,\text{for}\,x_{0.05}$
$P = 0.63\,\text{for}\,x_{0.5}$
$P = 0.185\,\text{for}\,x_{0.95}$

We can thus construct a nine-point logic tree (using three states for each of the variables) as shown below:

$P(q_o)$	q_o	$P(a)$	a	P	q	Sorted q	P	Cum P
0.185	568	0.185	0.0672	0.034225	290.0698	150.5238	0.034225	0.034225
0.63	650	0.185	0.0672	0.11655	331.946	172.2544	0.11655	0.150775
0.185	732	0.185	0.0672	0.034225	373.8223	193.9849	0.034225	0.185
0.185	568	0.63	0.1	0.11655	208.9555	208.9555	0.11655	0.30155
0.63	650	0.63	0.1	0.3969	239.1216	239.1216	0.3969	0.69845
0.185	732	0.63	0.1	0.11655	269.2878	269.2878	0.11655	0.815
0.185	568	0.185	0.1328	0.034225	150.5238	290.0698	0.034225	0.849225
0.63	650	0.185	0.1328	0.11655	172.2544	331.946	0.11655	0.965775
0.185	732	0.185	0.1328	0.034225	193.9849	373.8223	0.034225	1

Using the definitions of weighted mean and weighted variance from Chapter 2 (i.e., by combining the calculated scenario probability, P, and the corresponding outcome, q), we obtain

$$E[q] = 243.9\,\text{bbl/d}$$
$$V[q] = 2762.6$$
$$SD[q] = \sqrt{2762.6} = 52.6\,\text{bbl/d}$$

Note that these values are essentially identical to the PEM results but slightly different from the FOSM results. Now, the CDF can be readily constructed using the (*Sorted q*) versus (*Cum P*) columns as shown above. The logic tree CDF is generally presented in a stair-step manner to explicitly show that it is the outcome of a combination of discrete states for the different uncertain variables.

The nine-point discrete LTA CDF, expressed in terms of exceedance probabilities (i.e., $P^* = 1 - P$), is shown in Fig. 6.25, along with the corresponding MCS values generated using 5000 LHS samples as described in Mishra (1998). Also shown is the CDF for normal distributions with the mean and standard deviations as calculated by PEM and the nine-point discrete distribution from LTA. All methods are seen to produce essentially similar CDFs.

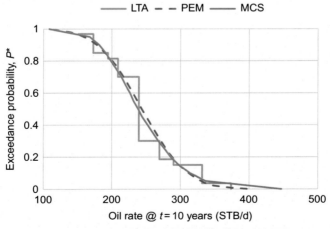

FIG. 6.25 Comparison of CDFs from PEM, LTA, FOSM, FORM, and MCS for the exponential decline Example 6.2.

6.6 TREATMENT OF MODEL UNCERTAINTY

6.6.1 Basic Concepts

The use of geostatistical techniques for generating multiple realizations of 3-D porosity and permeability fields to build static and dynamic reservoir models has become routine. Geostatistical methods can generate fine-scale images of reservoir properties honoring data from a variety of sources, with the realization-to-realization variations characterizing the uncertainty due to incomplete information and paucity of data. This uncertainty can be recognized as *conceptual or model uncertainty*, as opposed to *parameter uncertainty* that has been the focus of this chapter.

Quantifying the impacts of such uncertainty on forecasts of reservoir performance generally requires a statistical model averaging approach (Singh et al., 2010), using either formal Bayesian model averaging (e.g., Neuman, 2003) or a heuristic approach such as generalized likelihood uncertainty estimation (e.g., Beven and Binley, 1992), which would necessitate flow simulations for a large number of these plausible reservoir descriptions. However, computational constraints often preclude the use of a full suite of geostatistical models for reservoir forecasting. Typically, only a few selected realizations are used in detailed simulations to provide an indication of the range of uncertainty in reservoir performance. These realizations are selected via a ranking of stochastic reservoir models on the basis of some surrogate measure of reservoir performance. The ranking technique has generally become accepted as an economical way of quantifying the impact of uncertainty in reservoir description on reservoir performance (e.g., Ballin et al., 1992; Gómez-Hernández and Carrera, 1994).

The ranking methods described above typically provide estimates of reservoir performance corresponding to the "best" case, the "median" case, and the "worst" case. These qualifiers are obtained from the surrogate performance measure (e.g., volumetric sweep), and are also assumed to apply to the actual performance measure of interest (e.g., fractional water cut). However, summary statistics of reservoir performance (e.g., mean and standard deviation), which are needed for performing economic risk analysis, cannot be generated from these selected cases without knowing their likelihoods or weights. To that end, an efficient method for both ranking and weighting models, based on the logic tree analysis concepts, as proposed by Mishra et al. (2000) is discussed next.

6.6.2 Moment-Matching Weighting Method for Geostatistical Models

The idea behind this approach goes back to the work of Kaplan (1981) in the field of probabilistic seismic hazard and risk assessment, where logic trees are commonly used for propagating uncertainty. To avoid the problem of combinatorial explosion, which puts a practical limit on the number of uncertain variables, Kaplan suggested that a discrete distribution with multiple (e.g., 10 or greater) values should be replaced by a simpler one with 3–5 values to make the uncertainty analysis problem more tractable. The two distributions are made consistent by requiring that the weights (probabilities) for the new values be chosen so as to preserve the first few statistical moments. Mishra et al. (2000) have proposed using a discrete distribution of three values with moment-matching weights.

The first step is to decide which discrete values of the surrogate measure should be chosen for further analyses. In order to capture the full range of uncertainty, the median (50th percentile) value along with the 5th percentile at the low end and the 95th percentile at the high end are reasonable choices. The geostatistical models (realizations) corresponding to these discrete values then become candidates for carrying out detailed simulations.

The second step is to decide how to weight the simulation results from each of these selected realizations. The weighting methodology is based on the fact that any continuous distribution can be approximated by a discrete distribution such that the statistical moments of the original distribution are preserved. As depicted in Fig. **6.26**, this implies that if we chose the values x_1, x_2, and x_3 as discrete representations of the surrogate performance measure, x, then their respective weights, P_1, P_2, and P_3, must satisfy the following moment-matching constraints:

$$P_1*x_1 + P_2*x_2 + P_3*x_3 = E[x] \tag{6.29}$$

$$P_1*x_1{}^2 + P_2*x_2{}^2 + P_3*x_3{}^2 = E\left[x^2\right] = E^2[x] + V[x] \tag{6.30}$$

where $E[\bullet]$ denotes the statistical expectation or average and $V[\bullet]$ denotes the variance. Note that x is some readily computed surrogate measure (indicator) of reservoir performance and the values x_1, x_2 and x_3 correspond to realizations R_1, R_2, and R_3, respectively.

From the continuous distribution of x, we know $E[x]$ and $V[x]$. Thus, once the discrete quantities x_1, x_2, and x_3 are chosen, the weights P_1, P_2, and P_3 are determined using Eqs. (6.29) and (6.30), and an additional constraint requiring that the discrete weights (probabilities) must sum to unity:

$$P_1 + P_2 + P_3 = 1 \tag{6.31}$$

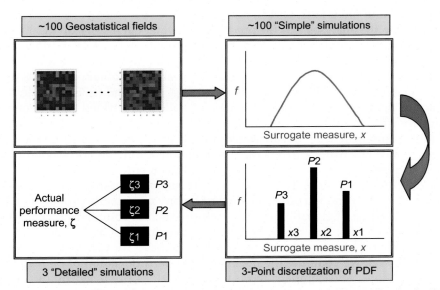

FIG. 6.26 **Schematic of proposed methodology, based on approximating a continuous distribution by a discrete distribution.**

The realizations selected in this process (R_1, R_2, and R_3) are then used as inputs to detailed simulations for computing the actual performance measure of interest (e.g., water cut and oil recovery) denoted by ζ. The uncertainty in forecasts of ζ can be characterized as follows:

$$M[\zeta] = P_1^* \zeta_1 + P_2^* \zeta_2 + P_3^* \zeta_3 \tag{6.32}$$

$$SD[\zeta] = P_1^* \{\zeta_1 - M[\zeta]\}^2 + P_2^* \{\zeta_2 - M[\zeta]\}^2 \\ + P_3^* \{\zeta_3 - M[\zeta]\}^{21/2} \tag{6.33}$$

where $M[\bullet]$ denotes the mean and $SD[\bullet]$ denotes the standard deviation.

6.6.3 Example Field Application

Mishra et al. (2000) present an example application of this methodology using a three-dimensional field example from the North Robertson unit (NRU) in west Texas. NRU is a heterogeneous, low permeability carbonate reservoir with 144 active producing wells and 109 injection wells. A smaller subset of the field containing 27 producers and 15 injectors was chosen for this analysis.

Fifty realizations of the permeability field were generated using sequential Gaussian simulation based on well-log data from 30 wells. The volumetric sweep efficiency, E_v, at 5000 days based on the single-phase tracer time-of-flight connectivity was calculated using a streamline simulator as the surrogate performance measure. The geostatistical models (realizations) were then ranked based on the computed CDF for volumetric sweep efficiency. Three realizations, corresponding to the 5th, 50th, and 95th percentile values of E_v, were selected for further analyses. The weights corresponding to these realizations were calculated as per Eqs. (6.29)–(6.31) and are 0.1593, 0.6473, and 0.1934.

The next step in the analyses was the prediction of water cut history up to 5000 days for all 27 producers and the cumulative oil recovery, in the model domain. Waterflood simulations incorporating multiphase flow effects were carried out for the three selected realizations, and their results were combined using the weights listed above to obtain estimates of the mean and standard deviation of water cut history as per Eqs. (6.32) and (6.33). In order to evaluate the accuracy of the proposed weighting scheme, detailed waterflood simulations were also carried out for all 50 realizations to compute the "true" mean and standard deviations. Fig. **6.27** (left) compares "true" mean water cut history (from all 50 simulations) with the "calculated" water cut history (from three simulations) and the corresponding standard deviation for a representative well, showing good agreement.

Note that predicting uncertainty in a "local" performance measure such as water cut forecasts on a well-by-well basis is a rather severe test for any uncertainty propagation technique, especially one based on a "global" surrogate measure such as volumetric sweep. A more reasonable test would be to examine the behavior of an actual "global" performance measure such as field-wide oil recovery, which is typically the basis for carrying out economic risk analysis. To this end, we compute the mean and standard deviation of oil recovery history for all 50 realizations, and compare those with the values predicted by our approximate method (using the 5–50–95 percentile weights). As shown in Fig. **6.27** (right), good agreement is also obtained between the two sets of results.

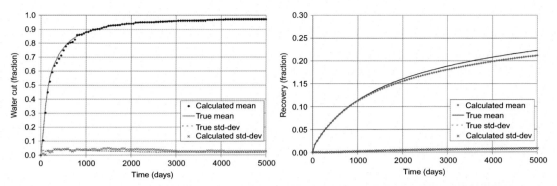

FIG. 6.27 Calculated mean and standard deviation of water cut response for well #6 (left) and field-wide oil recovery (right) using the 5–50–95 percentile weights as compared with the true mean and standard deviation calculated from all 50 realizations. *After Mishra, S., Choudhary, M.K., Datta-Gupta, A., 2000. A novel approach for reservoir forecasting under uncertainty. Soc. Pet. Eng. https://doi.org/10.2118/62926-MS.*

In summary, the moment-matching method provides a computationally expedient framework for dealing with model uncertainty as represented by multiple geostatistical realizations. It can calculate the mean and variance of reservoir performance while using only a few geostatistical models along with their weights computed on the basis of a surrogate performance measure.

6.7 ELEMENTS OF A GOOD UNCERTAINTY ANALYSIS STUDY

The following bulleted list provides some guidance regarding the desirable attributes of a good uncertainty analysis study. This list is generic and can be applied to quantitative uncertainty quantification studies from any problem domain.

- Define problem
  - Provide a succinct statement of problem including performance measures of interest.
  - Describe mathematical model linking inputs to outputs. For computationally expensive forward simulations, it might be necessary to build simplified or surrogate models using experimental design and response surface methods (see Chapter 7).

- Characterize input uncertainty
  - Show PDF/CDF in charts or tables; provide source of information.
  - If discrete states or expert judgment were used, describe how these were how obtained.
  - Discuss if there is any correlation among inputs and how this will be handled.

- Perform uncertainty propagation
  - Show PDF/CDF/risk profile of output.
  - Provide table of mean, SD, percentiles of interest.

    o Discuss statistical stability of results (i.e., sensitivity to sample size).
    o Discuss the application of simplified methods (e.g., FOSM), if appropriate.

- Determine sensitivity/importance ranking
  - o Show table/charts of importance ranking + scatterplots and describe what methods were used to obtain these.
  - o Provide tornado/spider (sensitivity) charts for comparison with MCS importance ranking.
  - o Discuss key drivers of risk and their physical implications from process understanding and future date collection perspectives.

- Summary and conclusions
  - o What is the range of likely outcomes and their probabilities?
  - o Which inputs are the key drivers of risk?
  - o What is the robustness of results to basic assumptions?
  - o What is the usefulness of probabilistic analysis and added value for decision-making?

6.8 SUMMARY

In this chapter, we started with the motivation for a probabilistic approach to uncertainty quantification. This was followed by a detailed discussion of each of element of our systematic framework (i.e., uncertainty characterization, uncertainty propagation, and uncertainty importance). A number of approaches were presented for each of these and demonstrated using several example problems. Finally, a practical methodology for treatment of model uncertainty was presented along with a field example.

Exercises

1. For the problem described in (6), calculate and plot spider and tornado charts. For a normal distribution, use mean ± 3 SD as the effective range.
2. For the porosity data given in Table 2.1 [POR_TAB-1.DAT], calculate the sampling distribution for the mean using the t-distribution and the normal approximation. Plot the CDFs. How do they compare? What is the reduction in variance going from the full sample distribution to the sample mean distribution?
3. Assume that the distribution of ϕ is U[0.1,0.3], while that of R is U[200,300]. Generate 10 samples for each of the variables using both random sampling and LHS. Prepare a scatter plot of ϕ versus R. Discuss the relative efficiencies of the two sampling techniques in covering the uncertain parameter space.
4. Consider Archie's equation for determining water saturation from a well log, viz.:

$$S_w^n = a\phi^{-m}/[R_t/R_w]$$

where S_w is water saturation, R_w is formation water resistivity, R_t is true formation resistivity, ϕ is porosity, and a, m, and n are empirical coefficients. Based on available information, we have:

$a = 0.62$

$n = 2$

$m = 0.33$

$\phi = N[0.20, 0.04]$

$R = R_t / R_w = U[250, 350]$

Calculate the CDF, $E[.]$ and $SD[.]$ of S_w using Monte Carlo simulation.

Using the *"random number generation"* option in *Excel* $\rightarrow$ *Tools* $\rightarrow$ *Data Analysis*, generate 100 samples of ϕ and R, and then combine them to compute the CDF of S_w (and its statistics). Your output should consist of: (a) a worksheet containing the random input vectors, and the corresponding output vector, (b) a worksheet showing the computation of the CDF, $E[.]$, and $SD[.]$, (c) a chart showing the CDF together with the position of the mean, (d) a chart showing a running tally of $E[.]$ and $SD[.]$, and (e) scatter plots of S_w versus ϕ and R. Which is the dominant input and why?

5. Create a contingency table between S_w and ϕ and S_w and R using 5×5 bins. Calculate the R-statistic for each case. Compare the significance with respect to the results in problem (4).

6. Volumetric estimates of oil in place can be computed using the formula:

$$N = 7758E - 6VS_o / B_o$$

where N is oil in place (MM STB), V is reservoir volume (ac-ft) $= N[70{,}000, 7000]$, S_o is oil saturation $(-) = U[0.50, 0.70]$, B_o is formation volume factor (RB/STB) $= T[1.15, 1.20, 1.25]$.

Calculate the mean and standard deviation for N using FOSM, PEM, and LTA. (Hint—convert U and T distributions to a 3-point distribution using the method used in Example 6.6.) How do the results compare? Calculate the fractional contribution to variance using FOSM based on Eq. (6.14).

7. Estimate the uncertainty in the permeability value calculated from the formula

$$k = \frac{162.6qB\mu}{mh}$$

where k is permeability (mD), q is flow rate (bbl/d) $= [240, 260]$, μ is viscosity (cp) $= 0.80$, B is formation volume factor $= 1.36$, m is slope of Horner plot (psi/log-cycle) $= 70 \pm 15$, h is thickness (ft) $= 69$.

Express your answer as percent relative error. Which parameter is the major source of uncertainty and why?

References

Arinkoola, A., Ogbe, D., 2015. Examination of experimental designs and response surface methods for uncertainty analysis of production forecast: a Niger delta case study. J. Pet. Eng.. 2015. https://doi.org/10.1155/2015/714541.

Ballin, P.R., Journel, A.G., Aziz, K.A., 1992. Prediction of uncertainty in reservoir performance forecasting. J. Can. Pet. Technol. 31, 52.

Beven, K.J., Binley, A., 1992. The future of distributed models: model calibration and uncertainty prediction. Hydrol. Process. 6, 279–298.

Bogen, K.T., 1994. A note on compounded conservatism. Risk Anal. 14, 379–381. https://doi.org/10.1111/j.1539-6924.1994.tb00255.x.

Bonnlander, B.V., Weigend, A.S., 1994. Selecting input variables using mutual information and nonparametric density estimation. In: Proceedings of the International Symposium on Artificial Neural Networks (ISANN'94), Tainan, Taiwan, pp. 42–50.

Bratvold, R.B., Begg, S., 2010. Making Good Decisions. Society of Petroleum Engineers, Richardson, TX.

Breiman, L., Friedman, J.H., Olshen, R.A., Stone, C.J., 1984. Classification and Regression Trees. Wadsworth and Brooks/Cole, Monterey, CA.

Caers, J., 2011. Modeling Uncertainty in the Earth Sciences. Wiley, New York.

Carreras, P.E., Johnson, S.G., Turner, S.E., 2006. Tahiti field: assessment of uncertainty in a deepwater reservoir using design of experiments. Soc. Pet. Eng. https://doi.org/10.2118/102988-MS.

Chang, C.-H., Yang, J.-C., Tung, Y.-K., 1997. Uncertainty analysis by point estimate methods. J. Hydraul. Eng.-ASCE 123 (3), 244–250.

Clemen, R.T., 1997. Making Hard Decisions. Duxbury, Pacific Grove, CA.

Dettinger, M.D., Wilson, J.L., 1981. First order analysis of uncertainty in numerical models of groundwater flow. Water Resour. Res. 17 (1), 149–157.

Draper, N.R., Smith, H., 1981. Applied Regression Analysis. John Wiley, New York.

Gómez-Hernández, J.J., Carrera, J., 1994. Using linear approximations to rank realizations in ground water modeling: application to worst case selection. Water Resour. Res. 30, 2065.

Granger, C.W.J., Lin, J., 1994. Using mutual information to identify lags in nonlinear models. J. Time Ser. 15, 371–384.

Hahn, G.J., Shapiro, S.S., 1967. Statistical Models in Engineering. John Wiley, New York.

Harr, M.E., 1987. Reliability-Based Design in Civil Engineering. McGraw-Hill, New York.

Harr, M.E., 1989. Probabilistic estimates for multivariate analyses. Appl. Math. Model. 13 (5), 313–318.

Hastie, T., Tibshirani, R., Friedman, J.H., 2008. The Elements of Statistical Learning: Data Mining, Inference, and Prediction. Springer, New York.

Helton, J.C., 1993. Uncertainty and sensitivity analysis techniques for use in performance assessment for radioactive waste disposal. Reliab. Eng. Syst. Saf. 42, 327–373.

Helton, J.C., Garner, J.W., McCurley, R.D., Rudeen, D.K., 1991. Sensitivity Analysis Techniques and Results for Performance Assessment at the Waste Isolation Pilot Plant, Report SAND9-7013. Sandia National Laboratories, Albuquerque, NM.

Hill, M.C., Tiedeman, C., 2007. Effective Ground water Model Calibration. Wiley-Interscience, Hoboken, NJ.

Hirasaki, G.J., 1975. Sensitivity coefficients for history matching oil displacement processes. Soc. Pet. Eng. https://doi.org/10.2118/4283-PA.

Iman, R.L., Conover, W.J., 1982. A distribution free approach to inducing rank correlation among inputs. Communications in Stats. Simul. Comput. 11, 335–360.

Iman, R.L., Conover, W.J., 1983. A Modern Approach to Statistics. John Wiley and Sons, New York, NY.

Iman, R.L., Helton, J.C., 1985. Investigation of uncertainty and sensitivity analysis techniques for computer models. Risk Anal. 8 (1), 71–90.

IPCC, 2010. Guidance Note for Lead Authors of the IPC Fifth Assessment Report on Consistent Treatment of Uncertainties, Inter Governmental Panel on Climate Change, accessed at https://www.ipcc.ch/pdf/supporting-material/uncertainty-guidance-note.pdf.

James, A.L., Oldenburg, C.M., 1997. Linear and Monte Carlo uncertainty analysis for subsurface multiphase contaminant transport. Water Resour. Res. 33 (11), 2495–2503.

Kaplan, S., 1981. On the method of discrete probability distributions. Risk Anal. 1, 189.

Keeny, R.L., von Winterfeld, D., 1991. Eliciting probabilities from experts in complex technical problems. IEEE Trans. Eng. Manag. 38 (3), 191–201.

Kessler, J.H., McGuire, R.M., 1999. Total system performance assessment using a logic tree approach. Risk Anal. 19, 915–932.

Ma, Y.Z., LaPointe, P. (Eds.), 2010. Uncertainty Analysis and Reservoir Modeling. AAPG Memoir 96.

MacDonald, R.C., Campbell, J.E., 1986. Valuation of supplemental and enhanced oil recovery projects with risk analysis. Soc. Pet. Eng.. https://doi.org/10.2118/11303-PAA.

McKay, M.D., Conover, W.J., Beckman, R.J., 1979. A comparison of three methods for selecting values of input variables in the analysis of output from a computer code. Technometrics 21 (3), 239–245.

Mishra, S., 1998. Alternatives to Monte-Carlo simulation for probabilistic reserves estimation and production forecasting. In: Presented at the SPE Annual Technical Conference and Exhibition, New Orleans, LA, 27–30 September 1998. https://doi.org/10.2118/49313-MS SPE-49313-MS.

Mishra, S., 2000. Uncertainty propagation using the point estimate method. In: Stauffer, F., Kinzelbach, W., Kovar, K., Hoehn, E. (Eds.), Calibration and Reliability in Groundwater Modeling: Coping with Uncertainty. In: IAHS Publication No. 265, International Association of Hydrological Sciences, Wallingford, pp. 292–296.

Mishra, S., 2002. Assigning Probability Distributions to Input Parameters of Performance Assessment Models. Report SKB-TR-02-11, Swedish Nuclear Fuel and Waste Management Co, Stockholm. 49 pp.

Mishra, S., 2009. Uncertainty and sensitivity analysis techniques for hydrologic modeling. J. Hydroinf. 11 (3–4), 282–296.

Mishra, S., Knowlton, R.G., 2003. Testing for input-output dependence in performance assessment models. In: Proceedings of the 10th International High-Level Radioactive Waste Management Conference, Las Vegas, NV.

Mishra, S., Parker, J.C., 1989. Effects of parameter uncertainty on predictions of unsaturated flow. J. Hydrol. 108, 19–25.

Mishra, S., Choudhary, M.K., Datta-Gupta, A., 2000. A novel approach for reservoir forecasting under uncertainty. Soc. Pet. Eng. https://doi.org/10.2118/62926-MS.

Mishra, S., Deeds, N.E., RamaRao, B.S., 2003. Application of classification trees in the sensitivity analysis of probabilistic model results. Reliab. Eng. Syst. Saf. 73, 123–129.

Mishra, S., Deeds, N., Ruskauff, G., 2009. Global sensitivity analysis techniques for probabilistic ground water modeling. Ground Water 47, 727–744. https://doi.org/10.1111/j.1745-6584.2009.00604.x.

Morgan, M.G., Henrion, M., 1990. Uncertainty: A Guide to Dealing with Uncertainty in Quantitative Risk and Policy Analysis. Cambridge University Press, New York.

Murtha, J.A., 1994. Incorporating historical data into Monte Carlo simulation. Soc. Pet. Eng. https://doi.org/10.2118/26245-PA.

Neuman, S.P., 2003. Maximum likelihood Bayesian averaging of uncertain model predictions. Stochastic Environ. Res. Risk Assess. 17 (5), 291–305.

Ovreberg, O., Damaleth, E., Haldorsen, H.H., 1992. Putting error bars on reservoir engineering forecasts. J. Pet. Technol. 44 (6), 732–738. https://doi.org/10.2118/20512-PA. SPE-20512-PA.

Press, W.H., Teuklosky, S.A., Vetterling, W.T., Flannery, B.P., 1992. Numerical Recipes in Fortran. Cambridge University Press, London.

RamaRao, B.S., Mishra, S., Andrews, R.W., 1998. Uncertainty importance of correlated variables in a probabilistic performance assessment. In: Proceedings of the SAMO'98, Second International Symposium on Sensitivity Analysis for Model Output, Venice, Italy, April 19–22.

Ravi Ganesh, P., Mishra, S., 2016. Simplified physics model of CO_2 plume extent in stratified aquifer-caprock systems. Greenhouse Gas Sci. Technol. 6, 70–82. https://doi.org/10.1002/ghg.1537.

Saltelli, A., Chan, K., Scott, M. (Eds.), 2000. Sensitivity Analysis. John Wiley, London.

Shannon, C.E., 1948. A mathematical theory of communication. Bell Syst. Tech. J. 27, 379–423.

Singh, A., Mishra, S., Ruskauff, G., 2010. Model averaging techniques for quantifying conceptual model uncertainty. Ground Water 48, 701–715. https://doi.org/10.1111/j.1745-6584.2009.00642.x.

Tung, Y.-K., Yen, B.-C., 2005. Hydrosystems Engineering Uncertainty Analysis. McGraw Hill Civil Engineering Series, New York.

Unlu, K., Parker, J.C., Chong, P.K., 1995. Comparison of three uncertainty analysis methods to assess impacts on groundwater of constituents leached from land-disposed waste. Hydrogeol. J. 3 (2), 4–18.

Venables, W.N., Ripley, B.D., 1997. Modern Applied Statistics with S-PLUS. Springer-Verlag, New York.

Walstrom, J.E., Mueller, T.D., McFarlane, R.C., 1967. Evaluating uncertainty in engineering calculations. Soc. Pet. Eng.. https://doi.org/10.2118/1928-PA.

Experimental Design and Response Surface Analysis

Numerical models are widely used in engineering and scientific studies. Making a number of simulation runs at various input configurations is what we call a computer experiment. The design problem is the choice of inputs for efficient analysis of data. Experimental design is an intelligent way to pick the choice of input combinations for minimizing the number of computer model runs for the purpose of data analysis, inversion problems, and input uncertainty assessment (Yeten et al., 2005; Schuetter and Mishra, 2014). One way to carry the tasks on experimental design results is to build a response surface. A response surface is an empirical fit of computed responses as a function of input parameters. In this chapter, we introduce various techniques for experimental design and response surface modeling and illustrate their application in petroleum engineering.

7.1 GENERAL CONCEPTS

To understand the behavior of a response function with respect to multiple predictor values, one typically needs a large number of observations to adequately cover the input space. An inefficient approach is to compute the response for all combinations of predictor values chosen on a suitably fine grid. Usually, this is not feasible. In physical experiments, some combinations of predictors may not be available to the experimenter or may produce responses that are beyond the capability of the instrumentation to measure. In numerically simulated experiments (e.g., finite-element- or finite-difference-based computer models), a large amount of computation may be required to collect each response. Therefore, computing responses over a grid of predictor values may take too long or be too expensive to complete.

The standard method for avoiding costly data collection is to only observe the response at prescribed combinations of predictors, called a design matrix, and then fit a metamodel (also called a "proxy model" or "response surface model" or "reduced-order model") to those points. These combinations are specially chosen to be representative of all possible predictor settings, called the input space. The runs are also chosen to allow the estimation of large-scale effects in the response. Using the observed runs, a statistical model is then developed. This model describes a specific mathematical relationship between the predictor variables and the response.

A good metamodel needs to have two characteristics. First, it must provide an accurate approximation of the full-physics simulation. That is, for any combination of predictor settings, the metamodel should predict a value of the response that is close to the value one would get by running the full simulation at the same settings. Second, the metamodel must run orders of magnitude faster than the full-physics simulation. If these two requirements are met, then the metamodel may be used as a proxy for the full-physics simulation, and since it can produce responses quickly, it can be used to explore the input space for optimal predictor combinations. In the petroleum and geoscience literature, some common applications of metamodels have been for model calibration or history matching (Li and Friedmann, 2007), parameter sensitivity analysis (White et al., 2001), uncertainty assessment using Monte Carlo methods (Friedmann et al., 2001; Carreras et al., 2006), reservoir studies (White and Royer, 2003; Ghomian et al., 2008), and optimal reservoir management (Esmaiel, 2005).

7.2 EXPERIMENTAL DESIGN

In this section, we introduce two broad categories of the design of experiments: factorial design and sampling design. Within each category, there are several choices. We discuss the pros and cons of these methods in terms of the number of simulation requirements and their space-filling characteristics.

7.2.1 Factorial Designs

Factorial designs are typically used for variable screening or response surface optimization. These designs set each of the predictor variables at one of several levels, usually a "low" and "high" or a "low," "center," and "high." Typically, "low," "center," and "high" levels are denoted −1, 0, and +1, respectively. When the number of inputs is small, factorial designs can use a relatively small number of runs to explore the predictor space and allow the estimation of simple linear or quadratic models, which can in turn be used to identify the regions of the space corresponding to optimal response values. As long as the response surface can be adequately modeled with simple functions, factorial designs are sufficient; however, other designs may be necessary for understanding the behavior of more complex functions (see Section 7.3). As the number of inputs increases, full-factorial designs can get quite large due to exponential growth in the number of runs. In that case, smaller factorial designs can be used to understand the response surface. A description of several of those designs is given below.

Plackett-Burman

Plackett-Burman designs (Plackett and Burman, 1946) are a class of designs that are chosen to provide the best possible estimates of the main effects of the predictors on the response. Main-effect estimates for Plackett-Burman designs have the minimum variance possible for a limited number of runs. The designs themselves are chosen so that each unique combination of levels for every pair of predictors appears the same number of times throughout the design. Typically, there are only two levels (+1 and −1) assigned for each input. While main effects are estimable, interaction effects between predictors are typically confounded with the main effects and cannot be separated without additional runs. Plackett-Burman designs for k inputs can have a number of unique runs anywhere between the nearest multiple of four from k (not any larger than $k+4$) and 2^k runs, where they become full 2^k factorial designs. One example of a Plackett-Burman design is shown in Fig. 7.1. In this case, the design has 12 runs

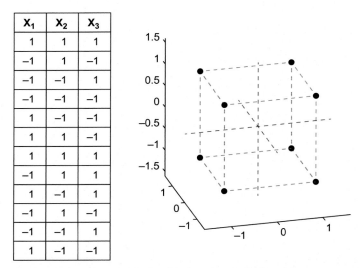

X_1	X_2	X_3
1	1	1
−1	1	−1
−1	−1	1
−1	−1	−1
1	−1	−1
1	1	−1
1	1	1
−1	1	1
1	−1	1
−1	1	−1
−1	−1	1
1	−1	−1

FIG. 7.1   **An example of a Plackett-Burman design for three inputs (left) and its representation in the predictor space (right).**

over 3 inputs, although there are only $2^3 = 8$ unique runs; the other runs are replicates. While replicates are commonly used in physical experiments to explore sources of variability, they do not impact the number of computer simulations.

Central Composite and Box-Behnken

Central composite (CC) and Box-Behnken (BB) (Box and Behnken, 1960) designs are related methods that use three levels for each predictor. Both designs make judicious use of observations and allow the estimation of linear and quadratic terms in a polynomial surface model. The CC design samples points at the corners of a hypercube in the input space and at points at the centers of the faces, as shown in Fig. 7.2. In contrast, the BB design samples points along the edges of the hypercube, as shown in Fig. 7.3. One commonly cited disadvantage to the CC design is that combinations where multiple predictors have simultaneous extreme values (i.e., at the corners of the hypercube) are typically unrealistic. The BB design places observations at less extreme predictor combinations to provide a better model fit over the center of the space.

Augmented Pairs

The augmented-pair (AP) design described by Morris (2000) is an alternative to central composite and Box-Behnken designs and is made to work well with sequential response surface search and optimization procedures. The strength of the AP design is that it builds the three-level targeted design by augmenting the two-level design used in the initial exploration phase. In this way, none of the runs are wasted. To construct an AP design, one begins with a two-level (preferably orthogonal) design, with observations at various combinations of -1 and $+1$ for the different factors. An example of such a design is the Plackett-Burman design. To augment the design, first, no center-point replicates are added (e.g., repeated runs with level 0 for all factors). Next, each pair of runs in the two-level design are used to construct

X$_1$	X$_2$	X$_3$
−1	−1	−1
−1	−1	+1
−1	+1	−1
−1	+1	+1
−1.68	0	0
0	−1.68	0
0	0	−1.68
0	0	0
0	0	1.68
0	1.68	0
1.68	0	0
+1	−1	−1
+1	−1	+1
+1	+1	−1
+1	+1	+1

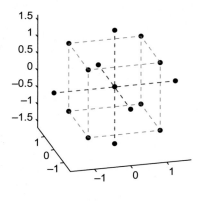

FIG. 7.2   **Central composite design for three inputs (left) and its representation in the input space (right). Note that the geometry of the design is specified by a parameter α that is set to be 1.68 for this specific case (rotatable CCD).**

X₁	X₂	X₃
−1	−1	0
−1	0	−1
−1	0	+1
−1	+1	0
0	−1	−1
0	−1	+1
0	0	0
0	+1	−1
0	+1	+1
+1	−1	0
+1	0	−1
+1	0	+1
+1	+1	0

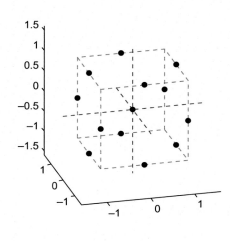

FIG. 7.3 **Box-Behnken design for three inputs (left) and its representation in the predictor space (right).**

X₁	X₂	X₃
0	0	0
1	1	1
−1	1	−1
1	−1	−1
−1	−1	1
0	1	0
1	0	0
0	0	1
0	0	−1
−1	0	0
0	−1	0

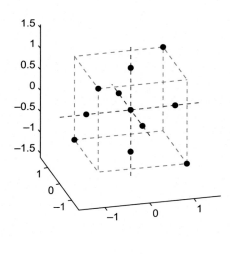

FIG. 7.4 **Augmented-pair design for three inputs (left) and its representation in the predictor space (right).**

a new single run, where the levels of the factors in the new run are $L_{new} = -0.5^*(L_1 + L_2)$. Here, L_1 and L_2 are the factor levels in the two parent runs, so that the new level of the factor will be 0 if the original runs were at +1 and −1, −1 if both original runs were at +1, or +1 if both original runs were at −1. The resulting design is smaller in size than a CC or BB design but still retains many of their advantages (Fig. 7.4).

Comparison of Factorial Designs

Fig. 7.5 shows a comparison of the number of unique runs required by each type of factorial design described above. The most expensive design is a full two-level factorial design, which has 2^k runs for k inputs (see the curve indicated in magenta). Such designs are a special case of Plackett-Burman design, but Plackett-Burman designs can have as low as $k+1$ runs. The minimum number of runs for a Plackett-Burman design is shown in Fig. 7.5 in cyan. Note, however, that such designs do not allow the estimation of much more than the main effects of the inputs and are not good in general for response surface modeling. Of the three-level designs, the Box-Behnken and central composite designs (red and green, respectively) have comparable numbers of unique runs, while the AP design typically has fewer runs. The maximum number of three-level runs possible is 3^k (not shown).

7.2.2 Sampling Designs

For smooth, well-behaved responses, factorial designs provide a means of fitting polynomial surfaces (e.g., linear for two-level designs and quadratic for three-level designs) to the data to guide further exploration in the predictor space. Because they were developed in the tradition of modeling physical experiments, predictors in these designs are only set to one of a few levels in each run; this allows the estimation of predictor effects (i.e., through an ANOVA decomposition) and the magnitude of the random variability present in the system. In this case, the goal is to fit a metamodel to the output of deterministic simulation code.

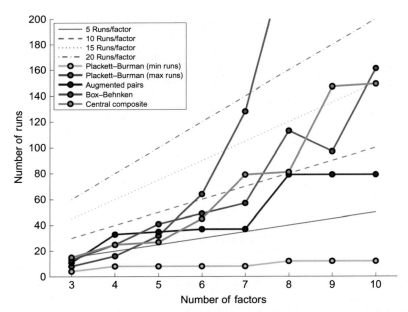

FIG. 7.5   **A comparison of the number of unique runs needed for the different factorial designs described in this section.**

That is to say, the variability in the system is zero. There is less of a need to sample predictors at one of a small set of values from run to run, since estimating variability is no longer required. Furthermore, it is possible that the simulation surface is not smooth and well behaved. There could be local discontinuities present that cannot be easily observed from a factorial design that only examines behavior at the low, center, and high end of the ranges for each predictor.

An alternative approach is a sampling design, which has runs that are not restricted to low, center, and high values of each predictor. Instead, the samples are randomly chosen across the ranges of values for each predictor. Generally, the goal is to have a space-filling design, that is, to spread observations across the predictor space with as few "holes" or "gaps" as possible.

Purely Random Design

The most basic sampling design is a simple random sample over the input space. Observations are chosen by drawing independent random samples of size n over the range of possible values for each input. The result is a design with n runs. Variations on this approach could use different marginal distributions in the sampling of the inputs or possibly include draws from a joint distribution over subsets of inputs. Random designs are easy and straightforward to produce. However, they could also suffer from poor "space-filling" characteristics. That is, multiple observations frequently end up clustered in one part of the space and provide largely redundant information about the behavior of the response surface in that region. Other parts of the space may be sparsely populated, and the redundant observations could be put to better use filling in those gaps.

Latin Hypercube Sampling

A Latin hypercube sample (LHS) design described by McKay et al. (1979) is intended to fill the predictor space by randomly selecting observations in equal probability bins across the range of the inputs. These designs sample values in [0, 1] for each of the inputs at each design point. The sampling is done in such a way that for a sample of size n, there will be exactly one observation in each of the intervals $[0, 1/n], [1/n, 2/n], \ldots, [(n-1)/n, 1]$ for each of the inputs.

In practice, the [0, 1] bounds on the values in LHS samples are interpreted to be a probability, and the design points are transformed through some probability distribution on the inputs. This has the effect of spreading the sampled points across equal regions of probability for each input, according to the chosen distribution. Several examples of LHS designs are shown in Fig. 7.6 for two predictors.

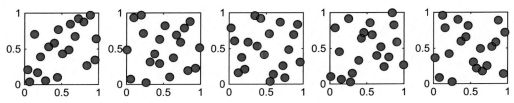

FIG. 7.6   **Examples of LHS designs using 20 observations for two predictors.**

Maximin LHS

A maximin LHS design described by Johnson et al. (1990) is created by generating a large number (e.g., thousands) of LHS designs and selecting the design that has the largest value of the function:

$$M(x^1, x^2, \ldots, x^n) = \min_{i,j} \|x^i - x^j\|,$$

where $x^1, x^2, \ldots, x^n$ are the n sampled observations and $\|x^i - x^j\|$ is the Euclidean distance between observations i and j. In other words, the maximin LHS design is the one that maximizes the minimum distance between any pair of observations in the sample. Examples of maximin LHS designs are shown in Fig. 7.7.

Maximizing the minimum distance between any pair of points has the effect of spreading the observations out as much as possible across the input space, under the constraint that the design is still based on a Latin hypercube. Maximin LHS designs, therefore, tend to have better space-filling characteristics. With a generic LHS design, there is a rare chance that, for example, all of the runs could be drawn from bins along the diagonal of the hypercube. This would result in a poor design for response surface modeling. Since maximin designs are selected from hundreds or thousands of candidate models, the chance of such a diagonal model is infinitesimally small. In general, for any location in the input space, the distance to the closest observation will be on average less in a maximin LHS design than in a generic LHS design.

Maximum Entropy Design

Maximum entropy designs described by Shewry and Wynn (1987) are also designed to have space-filling characteristics. The design is chosen to maximize the amount of "information" given by the sample, which in this case is captured by the entropy measure as defined in Shannon's information theory (Shannon, 2001). One way to do this is to maximize the determinant of the correlation matrix $\mathbf{C} = (r[i,j])$, where

$$r[i,j] = \begin{cases} 1 - \Gamma(h_{ij}) & \text{if } h_{ij} \leq a \\ 0 & \text{if } h_{ij} > a \end{cases}.$$

Here, h_{ij} is the distance between two observations x^i and x^j, and $\Gamma(h_{ij})$ is a spherical variogram with range a, defined by

$$\Gamma(h) = \frac{3h}{2a} - \frac{1}{2}\left(\frac{h}{a}\right)^3.$$

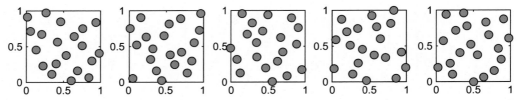

FIG. 7.7   Examples of maximin LHS designs using 20 observations for two predictors.

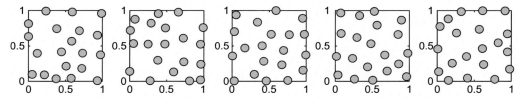

FIG. 7.8   Examples of maximum entropy designs using 20 observations for two predictors.

Maximum entropy designs are not restricted to equal probability bins, as LHS designs are. Several examples of these designs are shown in Fig. 7.8.

Comparison of Sampling Designs

The figures below show comparisons of the various types of sampling designs with respect to several space-filling criteria. The wraparound L_2 discrepancy described by Hickernell (1998), $WL2$, measures the difference between the number of design points per subvolume compared with the same count for a uniform distribution of points across the input space. It is computed with the formula shown below, where p is the number of inputs and $\mathbf{x}^1$, $\mathbf{x}^2, \ldots, \mathbf{x}^n$ are the n observations (i.e., design runs):

$$WL2 = -\left(\frac{4}{3}\right)^p + \frac{1}{n^2}\sum_{i=1}^{n}\sum_{j=1}^{n}\prod_{k=1}^{p}\left(\frac{3}{2} - \left|x_i^k - x_j^k\right|\left(1 - \left|x_i^k - x_j^k\right|\right)\right)$$

The second criterion is the maximin criterion:

$$M = \min_{i,j}\left\|\mathbf{x}^i - \mathbf{x}^j\right\|$$

The final criterion is the entropy measure, defined as $E = \det(\mathbf{C})$, where the matrix $\mathbf{C} = (r[i,j])$ as described in the maximum entropy design.

To compare the space-filling characteristics of each of the sampling designs, 100 designs of each type were sampled over $n = 20$ runs and $d = 2$ inputs. Each of the three criteria was then computed for each design. Comparisons of the designs are shown in Figs. 7.9–7.11,

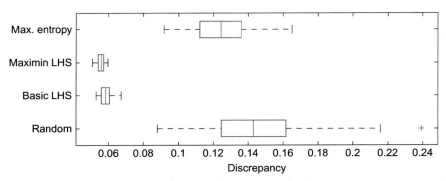

FIG. 7.9   Comparison of the sampling designs with respect to the wraparound L_2 discrepancy measure. Smaller values indicate better space-filling characteristics.

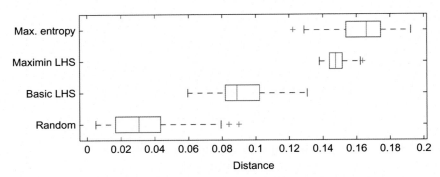

FIG. 7.10 **Comparison of the sampling designs with respect to the maximin distance measure. Larger values indicate better space-filling characteristics.**

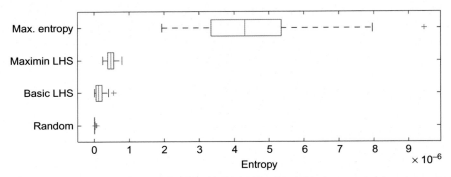

FIG. 7.11 **Comparison of the sampling designs with respect to the entropy measure. Larger values indicate better space-filling characteristics.**

corresponding to wraparound L_2 discrepancy, maximin, and entropy measures, respectively. Maximum entropy is the top performer for two of the metrics, including the maximin measure. It is able to outperform the maximin LHS design in the latter case because it is not bound by the restriction to be a Latin hypercube design. In terms of wraparound L_2 discrepancy, maximin LHS seems to outperform other designs.

7.3 METAMODELING TECHNIQUES

After deciding on an experimental design, the experiment can be run at each of the prescribed predictor settings, and the responses can be observed. Using the design and the observed response, a response surface model may be used to predict what the response would have been at an unobserved combination of predictor values. In the context of computer experiments, where the response being modeled is the result of deterministic computer code, the response surface model is also referred to as a proxy model or a metamodel. Both terms capture the fact that one is using a model (i.e., the metamodel) to predict the output of another model (i.e., the deterministic computer code).

There are many variations of metamodels, but the goal is generally the same for all of them. Some assumptions are made about either the shape of the response surface, its smoothness, and/or the correlation in responses between points that are close in the space. The parameters for these assumptions are estimated with the sampled observations, and a criterion is optimized. Typically, that criterion balances the smoothness and simplicity of the surface with its ability to match available data.

7.3.1 Quadratic Model

The quadratic polynomial model fits a parametric model to the response that is the analog of the parabola in p dimensions. It is defined as a sum of all linear, quadratic, and pairwise cross product terms between predictors. That is, the approximating function $\hat{f}(\mathbf{x})$ is defined by

$$\hat{f}(\mathbf{x}) = \hat{y} = b_0 + \sum_{i=1}^{p} b_i x_i + \sum_{i=1}^{p} b_{ii}(x_i)^2 + \sum_{i=1}^{p}\sum_{j>i} b_{ij} x_i x_j$$

The coefficients in the quadratic polynomial model are estimated by solving the linear model $\mathbf{Y} = \mathbf{XB}$, where

$$Y = \begin{pmatrix} f(x^1) \\ f(x^2) \\ \vdots \\ f(x^n) \end{pmatrix}, \quad X = \begin{pmatrix} 1 & x_1^1 & \cdots & x_p^1 & (x_1^1)^2 & \cdots & (x_p^1)^2 & x_1^1 x_2^1 & x_1^1 x_3^1 & \cdots & x_{p-1}^1 x_p^1 \\ 1 & x_1^2 & \cdots & x_p^2 & (x_1^2)^2 & \cdots & (x_p^2)^2 & x_1^2 x_2^2 & x_1^2 x_3^2 & \cdots & x_{p-1}^2 x_p^2 \\ \vdots & \vdots & & \vdots & \vdots & & \vdots & \vdots & \vdots & & \vdots \\ 1 & x_1^n & \cdots & x_p^n & (x_1^n)^2 & \cdots & (x_p^n)^2 & x_1^n x_2^n & x_1^n x_3^n & \cdots & x_{p-1}^n x_p^n \end{pmatrix},$$

and

$$\mathbf{B} = \left(b_0, b_1, \ldots, b_p, b_{11}, \ldots, b_{pp}, b_{12}, b_{13}, \ldots, b_{p-1,p}\right)^T.$$

The solution is given by $\hat{B} = (\mathbf{X'X})^{-1}\mathbf{X'Y}$. This is an example of multivariate linear regression discussed in Chapter 4.

7.3.2 Quadratic Model With LASSO Variable Selection

Typically, in industry, the analyst will perform a variable selection technique before proceeding with a quadratic fit. This could be done, for example, using exploratory analysis, stepwise regression, or comparison of candidate models using information criteria like AIC or BIC (Chapter 4). Ultimately, the final model fit will only use a subset of the main effects, interactions, and squared effects. This results in a parsimonious model and can often lead to better predictions because noisy, less relevant covariates have been removed from consideration.

One way of performing variable selection is through an automatic procedure based on LASSO regression. Least absolute shrinkage and selection operator (LASSO) regression described by Tibshirani (1996) is a technique for fitting a basic multiple linear regression model while shrinking the coefficients toward zero. Mathematically, this is done by adding a penalty term to the least-squares term in the objective function for linear regression (Eq. 4.18b).

$$\text{Minimize} \sum_i^n \left(Y_i - \hat{\beta}_0 - \sum_{j=1}^p \hat{\beta}_j X_{ij} \right)^2 + \lambda \sum_{j=1}^p |\beta_j|$$

LASSO regression has the interesting property that some of the fitted coefficients will be exactly zero. In these cases, LASSO serves as a variable selection algorithm where variables whose coefficients are zero are removed from the model.

The full procedure for the LASSO variable selection and quadratic fit is as follows:

(1) Determine an appropriate value of the strength of the penalty term λ, typically using cross validation on the root-mean-square error (RMSE) of the regression fit. Large values of λ will drive more coefficients toward zero.

(2) Fit a LASSO model using the quadratic regression model below by minimizing the least squares error with the penalty term:

$$\hat{f}(\mathbf{x}) = \hat{y} = b_0 + \sum_{i=1}^p b_i x_i + \sum_{i=1}^p b_{ii}(x_i)^2 + \sum_{i=1}^p \sum_{j>i} b_{ij} x_i x_j$$

(3) Identify which coefficients (b_0, b_i, b_{ij}, and b_{ii}) are nonzero in the LASSO model. Remove all main effects, interactions, and squared terms that are associated with the zero coefficients.

(4) Refit an ordinary least-squares regression model using only the remaining terms from the LASSO model.

7.3.3 Kriging Model

The kriging model described by Cressie (1993) and Krige (1951) has an approximation function that is composed of a trend term and an autocorrelation term. That is,

$$\hat{f}(\mathbf{x}) = \mu(\mathbf{x}) + Z(\mathbf{x}),$$

where $\mu(\mathbf{x})$ is the overall trend and $Z(\mathbf{x})$ is the autocorrelation term. $Z(\mathbf{x})$ is treated as the realization of a mean-zero stochastic process with a covariance structure given by $\text{Cov}(Z(\mathbf{x})) = \sigma^2 \mathbf{R}$, where $\mathbf{R}$ is an $n \times n$ matrix whose (i, j)th element is the correlation function $R(\mathbf{x}^i, \mathbf{x}^j)$ between any two of the sampled observations $\mathbf{x}^i$ and $\mathbf{x}^j$. *Ordinary kriging* assumes a scalar trend $(\mathbf{x}) = \mu_0$, whereas *universal kriging* uses a parametric trend term.

The Matérn correlation is often favored for kriging models because it tends to produce estimates that are smoother on a local level than other common alternative structures, like the exponential. However, it is also more flexible than Gaussian correlation, which can be overly smooth. An example of the Matérn $(5/2, \theta)$ correlation where $d_k = \left(x_k^i - x_k^j \right)$ is given below:

$$R(\mathbf{x}^i, \mathbf{x}^j) = \prod_{k=1}^p \left[1 + \frac{d_k \sqrt{5}}{\theta_k} + \frac{5 d_k^2}{\theta_k^2} \right] \exp\left(-\frac{d_k \sqrt{5}}{\theta_k} \right)$$

In the universal kriging model, the quadratic polynomial trend term below is commonly used:

$$\mu(\mathbf{x}) = b_0 + \sum_{i=1}^{p} b_i x_i + \sum_{i=1}^{p} b_{ii}(x_i)^2 + \sum_{i=1}^{p}\sum_{j>i}^{p} b_{ij} x_i x_j$$

7.3.4 Radial Basis Functions

Radial basis functions described by Chen et al. (1991) are any functions that depend solely on the distance of an observation to some fixed location $\mathbf{c}$. That is, an RBF $\phi(\cdot)$ satisfies $\phi(\mathbf{x}) = \phi(\|\mathbf{x} - \mathbf{c}\|)$. An RBF regression model takes the following form:

$$\hat{f}(\mathbf{x}) = b_0 + \sum_{i=1}^{p} b_i \phi_i(\|\mathbf{x} - \mathbf{x}_i\|)$$

That is, the response surface is approximated by a weighted sum of radial basis functions, each of which depends on the distance from the location of interest, $\mathbf{x}$, and one of the sampled observations, $\mathbf{x}_i$. The regression weights b_i are then trained using an ordinary least-squares approach. Other variations on this theme may be used to improve model fit. One way to provide a smoother fit is to include a smaller number of basis functions that involve alternative centers $\mathbf{c}_1, \mathbf{c}_2, \ldots, \mathbf{c}_{p'}$ instead of $\mathbf{x}_1, \mathbf{x}_2, \ldots, \mathbf{x}_p$, where $p' \ll p$. Another alternative is to allow the parameters of the $\phi_i(\cdot)$ functions to vary by location.

7.3.5 Metamodel Performance Evaluation Metric

The most desirable property of a metamodel is that it will provide the closest match between the prediction and the true response for future independent test data. When comparing different metamodels, it is useful to be able to capture the quality of the metamodel fit in a single statistic. There are many ways to do this, but two of the most common are RMSE and R^2. RMSE is defined as the square root of the average squared difference between predictions $\hat{y}_i = \hat{f}(\mathbf{x}^i)$ and true response values $y_i = f(\mathbf{x}^i)$ for a set of observations $\{\mathbf{x}^1, \mathbf{x}^2, \ldots, \mathbf{x}^n\}$:

$$\text{RMSE} = \sqrt{\frac{1}{n}\sum_{i=1}^{n}(y_i - \hat{y}_i)^2}$$

The RMSE can also be normalized by, for example, dividing it by the median observed response. This puts it on a similar scale regardless of the response, allowing for comparison of metamodel fits to different response surfaces:

$$\text{``Scaled'' RMSE} = \text{SRMSE} = \frac{\sqrt{\frac{1}{n}\sum_{i=1}^{n}(y_i - \hat{y}_i)^2}}{\text{median}\{y_1, y_2, \ldots, y_n\}}$$

Another metamodel accuracy measure is R^2, which is defined as the amount of variation in the response that is explained by the predictors. In a simple linear regression model, the R^2

statistic is the square of the correlation between the actual and predicted response values. For other models, a pseudo-R^2 statistic is typically used:

$$\text{Pseudo} - R^2 = R_p^2 = 1 - \frac{SS_{model}}{SS_{error}} = 1 - \frac{\sum_{i=1}^{n} (y_i - \hat{y}_i)^2}{\sum_{i=1}^{n} (y_i - \overline{y})^2}$$

Note that while R^2 in simple linear regression is always in $[0, 1]$, the pseudo-R^2 is in $[-\infty, 1]$. A negative pseudo-R^2 statistic means the model predicts the response worse than a flat model that predicts the mean observed response value everywhere in the predictor space.

7.4 AN ILLUSTRATION OF EXPERIMENTAL DESIGN AND RESPONSE SURFACE MODELING

We illustrate here the steps involved in experimental design and response surface analysis using an example involving the prediction of flowing bottom-hole pressure (BHP) at a well for a given time as function of three variables: permeability (PERM), porosity (POR), and skin factor (SKIN). The skin factor is a dimensionless quantity that quantifies the near-wellbore damage. Thus, in this example, our response variable is the BHP, and the three factors are PERM, POR, and SKIN.

In experimental design, several parameters are varied simultaneously according to a predefined pattern. The design here refers to a set of factor value combinations for which responses are measured as discussed before. For the design, the first step is to specify the number of levels and assign appropriate value to the factors for each level. We use here a three-level Box-Behnken design, and the variable ranges are shown in Table 7.1.

A Box-Behnken design requires a less number of experiments compared with a full-factorial design. For example, in this case, the design requires 16 experiments for the three factors, including four replicates at the factor center point (all factors assigned to their center-point values) (Table 7.2). Center-point replicates make the design more nearly orthogonal, which improves the precision of estimates of the response surface coefficients. Whereas the center-point replicates are common in experimental data collection to ensure repeatability, they are less common in computer experiments. Without the center-point replicates, the Box-Behnken design here will require 13 experiments as shown in Fig. 7.3.

The next step is to obtain the response for the combination of factors in Table 7.2. For field applications involving complex geology, this typically requires numerical simulation of the reservoir response. For this example, the BHP history for the center point is shown in Fig. 7.12.

TABLE 7.1 Predictor Variable Ranges for Experimental Design

		PERM (MD)	POR	SKIN
Low	−1	0.05	0.2	−2
Center	0	0.1	0.25	0
High	1	1	0.3	1

TABLE 7.2   **Box-Behnken Design with Three Factors and Four Center-point Replicates**

EXP #	PERM	POR	SKIN	BHP (psi)
1	1	0	−1	2884.4
2	0	1	−1	2129.4
3	−1	0	−1	1360.4
4	0	−1	1	1488.9
5	−1	1	0	596.93
6	0	0	0	1711.1
7	−1	−1	0	515.89
8	0	1	1	1529.4
9	0	0	0	1711.1
10	0	0	0	1711.1
11	0	−1	−1	2088.8
12	1	1	0	2847.6
13	1	0	1	2824.4
14	1	−1	0	2839.7
15	−1	0	1	160.52
16	0	0	0	1711.1

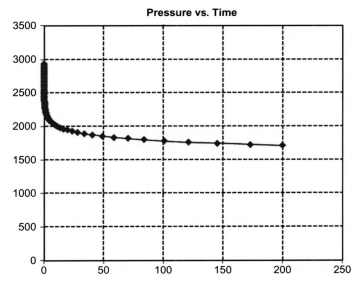

FIG. 7.12   **Bottom-hole pressure response for the combination of factors at the center point. The pressure at 200 days is used in the response surface analysis.**

The single response used in this example is the BHP at 200 days, although experimental design does allow for multiple responses to be modeled. The BHPs at 200 days for the various factor combinations in the Box-Behnken design are also included in Table 7.2.

The design step is followed by building of a response surface model that is an empirical fit of the response as a function of the factors. The Box-Behnken design is used to construct a second-degree polynomial response surface model given below:

$$BHP = b0 + b1*PERM + b2*POR + b3*SKIN + b4*PERM*PERM + b5*POR*POR + b6*SKIN*SKIN$$
$$+b7*PERM*POR + b8*PERM*SKIN + b9*POR*SKIN$$

The coefficients of the regression equation above are obtained by a multilinear regression, and the regression results are summarized in Table 7.3. The high value of R^2 indicates that most of the variability in BHPs can be explained by the regression model.

The results of analysis of variance for the multilinear regression are shown in Table 7.4. As discussed in Chapter 4, the analysis of variance (ANOVA) is actually a hypothesis test to examine the influence of the factors in explaining the response. The F-test statistic is used to test the hypothesis that none of the factors are linearly related to the response variable. A large value on the observed F-test statistic indicates that the linear model adequately explains the response. The P-value is defined as the probability of having a test statistic that is at least large as the observed test statistic. A small P-value means that at least some of the factors have effect on the response variable.

The next step is to examine the influence of the individual factor on the response. This is done by looking at the t-statistic associated with the regression coefficients. The results are summarized in Table 7.5. The first column indicates the coefficients in the polynomial model, and the second column shows its value. The third and fourth columns show the t-statistic and the associated P-significant values. Again, small P-values will indicate that the parameters are significant. Often, a threshold P-value, for example, P-value < 0.005, is set to test the significance. This means that the coefficients with P-value smaller than 0.005 are significant. The fifth column shows the standard error in the estimation of a particular response surface coefficient. The sixth and seventh columns show the -95% and $+95\%$ confidence values, respectively, for the coefficients.

TABLE 7.3   **Regression Summary**

R^2	0.998
R^2 adjusted	0.996
Standard error	56.48
# Points	16
R^2 for prediction	0.971

TABLE 7.4   **Analysis of Variance**

Source	df	SS	SS%	MS	F	P-Values
Regression	9	10725941.02	100	1191771.225	373.57	1.50861E−07
Residual	6	19141.1	0	3190.2		
Total	15	10745082.12	100			

TABLE 7.5   Test of Significance for the Coefficients

Coefficient	Coefficient Values	t-Stat	P-Value	Std Error	−95%	95%
b_0	1711.1	60.59	1.35841E−09	28.24	1642.0	1780.2
b_1	1095.3	54.85	2.46619E−09	19.97	1046.4	1144.2
b_2	21.26	1.064	0.328	19.97	−27.61	70.12
b_3	−307.48	−15.40	4.74339E−06	19.97	−356.34	−258.62
b_4	−6.382	−0.226	0.829	28.24	−75.48	62.72
b_5	−4.706	−0.167	0.873	28.24	−73.81	64.40
b_6	102.71	3.637	0.01088	28.24	33.61	171.81
b_7	−18.27	−0.647	0.542	28.24	−87.37	50.83
b_8	284.98	10.09	5.50004E−05	28.24	215.88	354.08
b_9	−2.5E−05	−8.85242E−07	1.000	28.24	−69.10	69.10

From Table 7.5, we can see relatively large P-values for the coefficients b_2, b_4, b_5, b_7, and b_9, indicating that these coefficients are likely to be zero. The results also indicate that porosity does not seem to influence the BHP that is consistent with our physical understanding of the problem. Repeating the multiple regression without these coefficients leads to the following equation for the response surface model:

$$\boxed{\text{Resp.1} = b0 + b1*perm + b2*skin + b3*perm*skin + b4*skin*skin}$$

The regression summary, the analysis of variance, and the test of significance table for the revised response surface model are given in Tables 7.6–7.8. The results show that there is no

TABLE 7.6   Regression Summary

R^2	0.998
R^2 adjusted	0.997
Standard error	47.04
# Points	16
R^2 for prediction	0.991

TABLE 7.7   Analysis of Variance

Source	df	SS	MS	F	P-Value
Regression	4	10720739.3	2680184.825	1211.1	1.84275E−14
Residual	11	24342.8	2213.0		
Total	15	10745082.12			

TABLE 7.8   Test of Significance for the Coefficients

		t-Stat	P-Value	Std Error	−95%	95%
b_0	1705.6	102.55	9.47447E−18	16.63	1669.0	1742.2
b_1	1095.3	65.85	1.22747E−15	16.63	1058.7	1131.9
b_2	−307.48	−18.49	1.24053E−09	16.63	−344.09	−270.87
b_3	284.98	12.12	1.05378E−07	23.52	233.21	336.75
b_4	102.71	4.367	0.00112	23.52	50.94	154.48

significant loss in R^2. The large F-values and the associated small P-value indicate that the model adequately explains the response. Finally, looking at the t-statistic and the corresponding P-values, we can conclude that all the coefficients in the response surface model are significantly different from zero.

Fig. 7.13 shows the cross plot of actual versus the predicted BHP using the response surface model. Clearly, the response surface model is able to predict BHP using the two factors: permeability and skin. Also, the residual plot (Fig. 7.14A) shows no clear structure, and the straight-line normal quantile plot for the residual (Fig. 7.14B) shows that the residuals are, indeed, normally distributed. These results seem to indicate that the response surface model does not violate the assumptions of the underlying regression model. Additional diagnostic plots with residual versus each of the factors can be made to further validate the model. Finally, Fig. 7.15 shows the response surface plot for various combinations of permeability and skin. As expected, we can see that as permeability decreases, the BHP also decreases. Similarly, as the skin factor increases, the BHP decreases. Again, these results are consistent with our physical understanding.

The reader can reproduce the results in this example using the public-domain software EREGRESS and the gas flow simulator GASSIM made available in the online resources for this book.

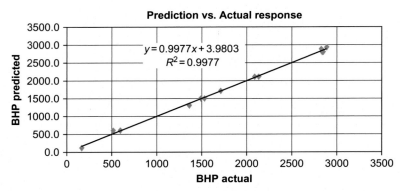

FIG. 7.13   Actual versus the predicted BHP using the response surface model.

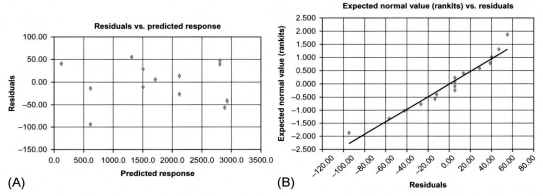

FIG. 7.14   **Diagnostic plots for the residuals: (A) residuals versus the fitted response and (B) normal quantile plot for the residuals.**

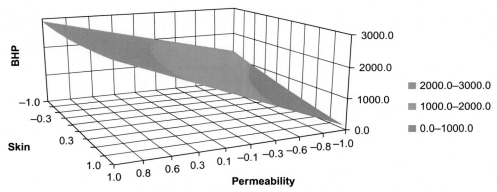

FIG. 7.15   **Plot of the response surface displaying the behavior of the BHP for various combinations of factors: permeability and skin.**

7.5 FIELD APPLICATION OF EXPERIMENTAL DESIGN AND RESPONSE SURFACE MODELING

7.5.1 Problem of Interest

There are numerous field applications of experimental design and response surface analysis for parameter sensitivity studies (White et al., 2001), fast surrogate modeling (Zubarev, 2009), geologic model calibration or history matching and uncertainty analysis (Cheng et al., 2008). We briefly discuss here an application for field-scale history matching of geologic models. History matching is the process of reconciling geologic models to dynamic reservoir response such as pressure data and multiphase production data. Effective strategies for history matching commonly follow a structured approach with a sequence of adjustments to the geologic model starting from global to regional parameters followed by local changes in model properties associated with matching for pressure, flood front progression, and individual well performance (Cheng et al., 2008; Yin et al., 2011). Typical parameters for pressure matching are regional pore volume multipliers, regional vertical and areal transmissibility multipliers, fault transmissibilities, and aquifer strength. Modern assisted/automatic history matching methods utilize design of experiments and response surface methodologies with machine learning and evolutionary algorithms to calibrate the uncertain parameters (Cheng et al., 2008).

7.5.2 Proxy Construction and Application Strategy

Key parameters are first identified via a sensitivity analysis and an initial ensemble of models that span the parameter ranges is created (Cheng et al., 2008). To minimize the number of flow simulations, which can be computationally demanding, it is common to construct surrogate or proxy models using experimental design and response surface analysis. The steps for sensitivity analysis and proxy modeling are illustrated in Fig. 7.16.

Genetic algorithm (GA), one of the evolutionary algorithms, is commonly used for model calibration (Cheng et al., 2008). The genetic algorithm imitates biological principles of evolution—survival of the fittest. The evolution starts from a population of randomly generated individuals consisting of a set of model parameters. In each generation, the fitness (a measure of dynamic data misfit) of every individual in the population is evaluated. A proxy model is particularly useful here because it can be used to reject individuals for

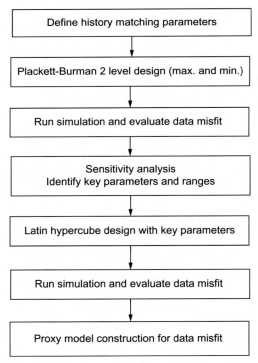

FIG. 7.16   **Flow chart of sensitivity analysis and proxy modeling for history matching.**

which the proxy approximation to the misfit function is higher than an acceptable threshold. This avoids costly flow simulations (Cheng et al., 2008; White and Royer, 2003; Yeten et al., 2005).

Multiple individuals are stochastically selected from the current population (based on their fitness) and modified (recombined and possibly randomly mutated) to form a new population. The new population is then used in the next iteration of the algorithm. The algorithm terminates when either a maximum number of generations have been produced, or a satisfactory fitness level has been reached for the population.

7.5.3 Field Case Study

We briefly discuss a field application of the genetic algorithm and proxy modeling for history matching (Yin et al., 2011). The **E** reservoir has an average depth of 1000 m with 13 active producers and six active injectors. The reservoir structural framework was built to represent seven zones. Each zone was divided into multiple layers resulting in a total of 424 layers. Also, 22 faults were incorporated into the model resulting in 12 fault blocks. The **E** field was operated under a combination of depletion and pressure maintenance strategy to maximize oil recovery and a better-calibrated full-field model would support these activities.

A stepwise illustration of the model calibration procedure is given in Fig. 7.17. To start with, a proxy model for the dynamic data misfit is constructed as discussed in Section 7.5.2. The proxy model is used to screen out less viable models with potentially large data misfit. This step results in substantial savings in computation time as only the models that pass the proxy screening are used for flow simulation and rigorous computation of the data

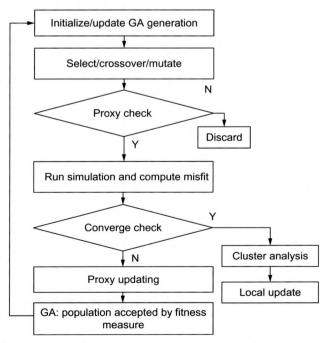

FIG. 7.17   **Flow chart of genetic algorithm with proxy for history matching.**

TABLE 7.9   E Field History Matching Parameters and Ranges

	Uncertainty Variables	Low Value	Mid Value	High Value	Distribution
Static uncertainties	Facies	−1	0	1	Discrete uniform
	Water saturation	−1	0	1	Discrete uniform
	Porosity	−1	0	1	Discrete uniform
Baffles/barriers	Interregion transmissibility multipliers	1.E−06	1.E−03	1.E+00	Continuous uniform
	Fault transmissibility multipliers	1.E−06	1.E−03	1.E+00	Continuous uniform
Relative permeability	Sorw1 (thin bed)	0.30	0.36	0.42	Continuous uniform
	Sorw2 (thick bed)	0.25	0.32	0.38	Continuous uniform
	Krwe1	0.30	0.45	0.60	Continuous uniform
	Krwe2	0.21	0.33	0.45	Continuous uniform
	Nw1	1.50	1.05	0.60	Continuous uniform
	Nw2	2.50	2.00	1.50	Continuous uniform
	Now1	3.20	11.60	20.00	Continuous uniform
	Now2	2.40	2.80	3.20	Continuous uniform
Rock property	Rock compressibility	5	20.50	36	Continuous uniform
Pore volume	Aquifer PV multiplier	1	10.50	20	Continuous uniform
Transmissibility	Horizontal transmissibility	0.5	0.75	1	Continuous uniform
	Vertical transmissibility	0.1	0.55	1	Continuous uniform

misfit. As shown in Fig. 7.17, these misfit calculations are used to further update the proxy model after sufficient number of flow simulations.

Table 7.9 shows a full set of uncertain parameters identified during geologic modeling. Given the large number of potential uncertain parameters, first a sensitivity analysis was carried out by a Plackett-Burman 2-level experimental design. Flow simulations were performed for each of the experiments and the effects of each parameter on the data misfit were ranked. The parameters resulting in the largest change of the misfit function were kept and the less sensitive parameters are discarded. The details of the sensitivity analysis can be found in Cheng et al. (2008).

An ensemble of geomodels was calibrated using GA with proxy model to history match well data including shut-in bottom-hole pressures, Modular Dynamic Test (MDT) pressures, and cumulative liquid productions. Some example matches of MDT pressure are shown in Fig. 7.18. A cluster analysis can be performed on the ensemble of history matched models

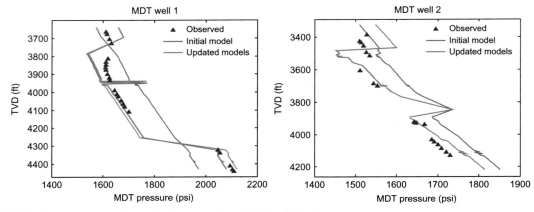

FIG. 7.18   Examples of Modular Dynamic Test (MDT) match.

to identify a diverse set of models for performance forecasting, uncertainty analysis and to optimize field development and management strategy (Cheng et al., 2008).

7.6 SUMMARY

In this chapter, we have introduced the concepts of experimental design and response surface analysis to construct proxy models to approximate the results of complex flow simulation. Proxy models can be very useful for large-scale field applications as they can be used to prescreen potential solutions without going through costly flow simulations. A good proxy can also substitute a full-physics simulation model within the prescribed range of predictor variables. Introduction of proxy models enables stochastic search and optimization algorithms such as GA, simulated annealing (SA), and Markov chain Monte Carlo (MCMC) to be practically feasible for field-scale applications, especially when there are large numbers of parameters in the problem. We have illustrated the power and utility of the experimental design and response surface models using a simple illustrative example and a field application.

Exercises

1. Suppose we have four inputs, with each input ranging between -1 and 1. Please create the following designs:

   (a) Full Plackett-Burman design (full factorial)
   (b) Central Composite design (with α is set to 1)
   (c) Box-Behnken design
   (d) Write a program to compute the Augmented Pairs' design, based on the 2-level full factorial design from (a)

2. Suppose we have three inputs. Each input has uniform probability and ranges between 0 and 1. Please finish the following:

   (a) Purely random design with 60 independent random samples.
   (b) Latin Hypercube Sampling with 60 observations.
   (c) Maximin LHS (among 50 realizations with 60 observations each realization).
   (d) Maximum entropy design (Hint: one option for sampling is to choose the one with highest entropy among 50 LHS realizations with 60 observations each realization).
   (e) Compare these designs using the three space-filling criteria mentioned in the book, with 100 realizations of designs for each type.

3. Consider the data given by Table 7.2. Ignore the porosity data (POR) and construct a second-degree polynomial response surface model.

   (a) What is the R^2 and standard error?
   (b) Provide the Analysis of Variance table.
   (c) Test the significance for the coefficients, regarding the constant term and PERM*PERM term. Is the conclusion changed compared to the results from Table 7.5?
   (d) Plot the diagnostic plots for the residuals. Is there any structural bias observed?

References

Box, G.E., Behnken, D., 1960. Some new three level designs for the study of quantitative variables. Technometrics 2 (4), 455–475.

Carreras, P.E., Johnson, S.G., Turner, S.E., 2006. Tahiti field: assessment of uncertainty in a deepwater reservoir using design of experiments. Soc. Pet. Eng. https://doi.org/10.2118/102988-MS.

Chen, S., Cowan, C.F., Grant, P.M., 1991. Orthogonal least squares learning algorithm for radial basis function networks. IEEE Trans. Neural Netw. 2 (2), 302–309.

Cheng, H., Dehghani, K., Billiter, T.C., 2008. A Structured Approach for Probabilistic-Assisted History Matching Using Evolutionary Algorithms: Tengiz Field Applications. Society of Petroleum Engineers. https://doi.org/10.2118/116212-MS.

Cressie, N., 1993. Statistics for Spatial Data (Wiley Series in Probability and Statistics). Wiley, New York.

Esmaiel, T.E., 2005. Applications of experimental design in reservoir management of smart wells. Soc. Pet. Eng. https://doi.org/10.2118/94838-MS.

Friedmann, F., Chawathe, A., Larue, D.K., 2001. Assessing Uncertainty in Channelized Reservoirs Using Experimental Designs. Society of Petroleum Engineers. https://doi.org/10.2118/71622-MS.

Ghomian, Y., Pope, G.A., Sepehrnoori, K., 2008. Development of a Response Surface Based Model for Minimum Miscibility Pressure (MMP) Correlation of CO2 Flooding. Society of Petroleum Engineers. https://doi.org/10.2118/116719-MS.

Hickernell, F., 1998. Lattice rules: how well do they measure up. In: Hellekalek, P., Larcher, G. (Eds.), Random and Quasi-Random Point Sets. Springer, New York, pp. 109–166.

Johnson, M.E., Moore, L.M., Ylvisaker, D., 1990. Minimax and maximin distance designs. J. Stat. Plan Inference 26 (2), 131–148.

Krige, D.G., 1951. A Statistical Approach to Some Mine Valuation and Allied Problems on the Witwatersrand. University of the Witwatersrand, Johannesburg.

Li, B., Friedmann, F., 2007. Semiautomatic multiple resolution design for history matching. Soc. Pet. Eng. https://doi.org/10.2118/102277-PA.

McKay, M.D., Beckman, R.J., Conover, W.J., 1979. Comparison of three methods for selecting values of input variables in the analysis of output from a computer code. Technometrics 21 (2), 239–245.

Morris, M.D., 2000. A class of three-level experimental designs for response surface modeling. Technometrics 42 (2), 111–121.

Plackett, R.L., Burman, J.P., 1946. The design of optimum multifactorial experiments. Biometrika 33, 305–325.

Schuetter, J., Mishra, S., 2014. Simplified predictive models for CO_2 sequestration performance assessment. Research topical report on statistical learning based models. Submitted to U.S. Department of Energy National Energy Technology Laboratory, October.

Shannon, C.E., 2001. A mathematical theory of communication. ACM SIGMOBILE Mob. Computi. Commun. Rev. 5 (1), 3–55.

Shewry, M.C., Wynn, H.P., 1987. Maximum entropy sampling. J. Appl. Stat. 14 (2), 165–170.

Tibshirani, R., 1996. Regression shrinkage and selection via the lasso. J. R. Stat. Soc. Ser. B Methodol. 58, 267–288.

White, C.D., Willis, B.J., Narayanan, K., Dutton, S.P., 2001. Identifying and Estimating Significant Geologic Parameters With Experimental Design. Society of Petroleum Engineers. https://doi.org/10.2118/74140-PA.

White, C.D., Royer, S.A., 2003. Experimental Design as a Framework for Reservoir Studies. Society of Petroleum Engineers. https://doi.org/10.2118/79676-MS.

Yeten, B., Castellini, A., Guyaguler, B., Chen, W.H., 2005. A Comparison Study on Experimental Design and Response Surface Methodologies. Society of Petroleum Engineers. https://doi.org/10.2118/93347-MS.

Yin, J., Park, H.-Y., Datta-Gupta, A., King, M.J., 2011. A hierarchical streamline-assisted history matching approach with global and local parameter updates. J. Pet. Sci. Eng. 80 (1), 116–130.

Zubarev, D.I., 2009. Pros and Cons of Applying Proxy-models as a Substitute for Full Reservoir Simulations. Society of Petroleum Engineers. https://doi.org/10.2118/124815-MS.

Further Reading

Banfield, J.D., Raftery, A.E., 1993. Model-based Gaussian and non-Gaussian clustering. Biometrics 49, 803.

Davis, J.C., 1986. Statistics and Data Analysis in Geology, second ed. John Wiley & Son, New York. p. 527.

Doveton, J.H., Prensky, S.E., 1992. Geological applications of wireline logs—a synopsis of developments and trends. Log Anal. 33, 286.

Hastie, T., Tibshirani, R., 1990. Generalized Additive Models. Chapman and Hall, London. p. 335.

Mahalanobis, P.C., 1936. On generalized distance in statistics. Proc. Natl. Inst. Sci. India 12, 49.

Mardia, K.V., Kent, J.T., Bibby, J.M., 1979. Multivariate Analysis. Academic Press, London. p. 521.

Serra, O., Abbott, H.T., 1982. The contribution of logging data to sedimentology and stratigraphy. SPEJ 22, 117.

The focus of this chapter is data-driven modeling, where machine-learning techniques are used to uncover the relationship between input and output variables. Our discussion will cover: (a) *premises*, i.e., easy-to-understand descriptions of the commonly used concepts and techniques, (b) *promises*, i.e., case studies demonstrating successful practical applications, and (c) *perils*, i.e., honest appraisal of challenges and potential pitfalls.

8.1 INTRODUCTION

8.1.1 Preliminaries

Big data analytics and data-driven modeling have become quite the buzzwords in recent years in the context of analyzing the performance of oil and gas reservoirs (Saputelli,

2016). Their growing application has been predicated on the potential to usher in exciting new developments related to (1) acquiring and managing data in large *volumes*, of different *varieties*, and at high *velocities* (the 3V problem) and (2) using statistical techniques to "mine" the data and discover hidden patterns of association and relationships in large, complex, multivariate datasets (Holdaway, 2014). The terms *data mining, statistical learning, knowledge discovery*, and *data analytics* have all been used interchangeably in this context. Essentially, the goal of such an exercise is to extract important patterns and trends and understand "what the data says," using supervised and/or unsupervised learning (Hastie et al., 2008).

In *supervised learning*, the value of an outcome is predicted based on a number of inputs, with the training dataset used to build a predictive model or "learner" via techniques such as regression analysis discussed in Chapter 4 and other methods to be discussed in this chapter. On the other hand, *unsupervised learning* involves describing associations/patterns among a set of input measures to understand how the data are organized or clustered, using techniques such as cluster analysis and principal component analysis discussed in Chapter 5 and other methods such as multidimensional scaling and self-organizing maps (see Hastie et al., 2008, for details).

8.1.2 Data-Driven Models—What and Why?

In classical statistics, the standard approach to data analysis requires postulating a model between the independent (predictor) and dependent (response) variables. For example, Chapter 4 discusses the application of linear regression for input-output modeling, where a simple linear relationship is assumed between the variables. As datasets have become more complex and/or multidimensional, there is a growing recognition that one needs to look beyond linear regression (or its linearizing variants) to better describe input-output relationships. In particular, the idea is to extract the model from the data without making any assumptions regarding the underlying functional form (Breiman, 2001b). This is what was also referred to earlier as supervised learning. Such problems can be further subdivided into (a) *regression* problems, where the response variable is continuous (e.g., permeability), or (b) *classification* problems, where the response variable is categorical (e.g., rock type). In both cases, the predictor variables can be continuous and/or categorical. For example, building a predictive model for the cumulative annual production in the first 12 months is a regression problem (Schuetter et al., 2015), whereas determining the factors responsible for identifying electrofacies on the basis of well-log response is a classification problem (Perez et al., 2005).

The benefits of data-driven modeling, compared with standard linear or nonlinear regression analysis with a prespecified data model, are (a) identifying hidden patterns in the data, (b) capturing complex nonlinear relationships between variables, (c) avoiding the need to explicitly define functional forms for the input-output relationship, (d) automatically handling correlations between predictors, and (e) guided/automated tuning of the model during "learning." However, some degree of model interpretability can be lost because of model complexity—leading to their labeling as "black-box" models.

8.1.3 Our Philosophy

For more than a decade, advanced algorithms developed by statisticians and computer scientists have been used to provide data-driven insights into system performance in fields ranging from consumer marketing to cyber security to health care (Bahga and Madisetti, 2016). Such techniques are also increasingly being used for problems such as reservoir characterization (e.g., Toth et al., 2013; Bhattacharya et al., 2016), production data analysis (e.g., Shelley et al., 2014; Lolon et al., 2016), reservoir management (e.g., Maučec et al., 2011; Maysami et al., 2013), and predictive maintenance (e.g., Rawi et al., 2010; Santos et al., 2015). However, the subject of data-driven modeling (*aka* data analytics) remains a mystery to most petroleum engineers and geoscientists because of the statistics-heavy jargon and the use of complex "black-box" algorithms.

Since the development or coding of advanced statistical algorithms is typically not the primary focus for petroleum engineers and geoscientists, there is increasing reliance on commercial packages such as *SAS* or open-source packages such as *R* that make these algorithms readily available to the larger community. Nonetheless, there still remains the issue of (a) choosing the right algorithm(s) for the problem as opposed to using a preferred one for all cases, (b) applying the algorithm(s) with the proper choice of user-defined parameters, (c) avoiding the problem of data overfitting and resulting bias in fitted model predictions, and (d) ensuring that the data-driven model makes physical sense in terms of variable selection and parameter importance.

In this chapter, we will provide an overview of some of the most commonly used data-driven modeling techniques (which can handle both regression and classification problems) for the petroleum geosciences. Since our focus will be on the application of the algorithms rather than their programming, the mathematical descriptions will be kept to a minimum. Our discussion will emphasize a thought process and analytic framework that can be easily applied by geoscientists and petroleum engineers, working together with data scientists. To that end, we will provide easy-to-understand descriptions of the algorithms supplemented by simple pedagogical examples and practical field examples demonstrating their applicability.

8.2 MODELING APPROACHES

8.2.1 Classification and Regression Trees

Classification and regression trees (CART) are simple, interpretive models to describe how the predictors impact the response (Breiman et al., 1984). The general idea is to (a) split the predictor space into nested rectangular regions, and (b) within each region, predict the response with a constant value for a regression problem (i.e., $y_i = c_i$) or a categorical label for a classification problem (i.e., $y_i = class_i$). As shown in Fig. 8.1, the resulting binary tree (right panel) is useful for determining prediction rules that partition output into groups based on input values and for finding structure in data. The method is called a tree, because the rectangular regions are defined by using a branching structure, and each branch is a binary split obtained by applying a threshold to the value of one of the predictors.

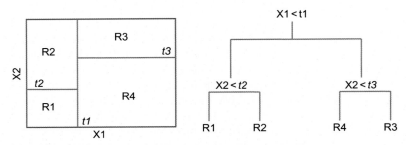

FIG. 8.1   **Schematic of tree-based modeling concept showing partitioning of parameter space into rectangular regions (left) and the corresponding binary tree (right).**

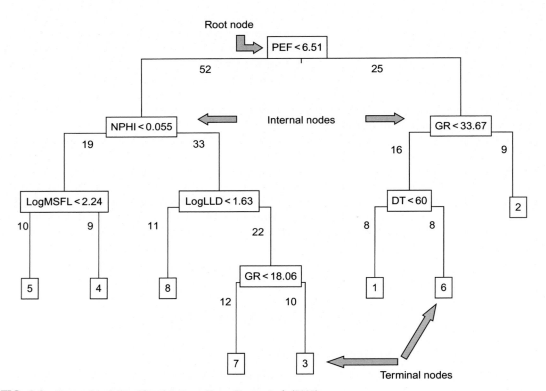

FIG. 8.2   **Example of classification tree. *From Perez et al. (2005).***

Fig. 8.2 shows an example of a *classification tree* that involves identification of electrofacies from multiple well-log responses for the Salt Creek field data described earlier in Chapters 4 and 5 (Perez et al., 2005). The tree has a root node where the binary splitting process starts and internal nodes where the splitting process continues until a terminal node is reached. The root node shows the first split or decision rule, PEF (photoelectric) < 6.51. With this split, the data of 77 samples are classified into two groups. The first group (52 samples) is placed to the left, and the second group (25 samples) is placed to the right in the tree. The next split corresponds to NPHI (neutron porosity) <0.055, where the 52 samples are divided into a group of 19 samples to the left and 33 samples to the right. If we continue this process with the left part of the tree, the final split is applied to logMSFL (micro spherically focused log) <2.24 resulting in two terminal nodes. The first split classified 10 samples as electrofacies 5 and the second split classifies 9 samples as electrofacies 4. From this figure, we can also easily deduce that the most important well logs for this classification problem are the ones near the top of the tree, that is, PEF, NPHI, and gamma ray (GR).

The CART construction requires the following parameters to be chosen at each split: predictor j, threshold values s, and predicted response c_1 and c_2 within each branch. For regression problems, c_1 and c_2 are the mean values of the response variable for each of the branches. For classification problems, they are the class labels corresponding to the category with the highest probability in each branch. Estimation of these parameters at each split requires minimizing some measure of misclassification error or node impurity (i.e., data misfit), such as a sum-of-squared error metric for regression or the Gini index (i.e., summation over product of class membership probability and its complement) for classification (Hastie et al., 2008). Once the best split is found, the data are partitioned into two mutually exclusive regions. The splitting process is then repeated on each of the two regions (and all the resulting regions) until the tree-building process is terminated.

The optimal tree size should be a parsimonious compromise between complexity of the tree and overall goodness of fit. One commonly used "pruning" process involves growing the tree to nearly full size and then selecting the subtree that optimizes some complexity criterion (Breiman et al., 1984). Generally, this is taken to include both a summation term representing overall node impurity and a penalty term combining a tuning (cost complexity) parameter and the number of terminal nodes. The cost-complexity parameter thus governs the trade-off between tree size and its goodness of fit to the data, with larger values of the parameter corresponding to smaller trees and vice versa. For example, the pruning chart shown in Fig. 8.3 displays this trade-off for the tree shown earlier, indicating that the misclassification error does not change significantly if we reduce the size of the tree from 39 to 25 nodes, with the biggest change occurring when there are fewer than 10 nodes (Perez et al., 2005). Additional computational details regarding the construction and pruning of regression and classification trees can be found in Hastie et al. (2008).

Once the optimal tree has been constructed, the most important predictors can be readily identified as the ones near the top of the tree. For example, in the classification problem illustrated in Fig. 8.2, the most important well logs are the photoelectric (PEF), neutron porosity (NPHI), and density (DT). If two or three variables are identified as holding most of the explanatory power in the model, the results can be visualized further through the use of a partition plot—especially for a classification problem. This is a scatterplot of the two most important input variables, with the categorical outcomes defined by unique symbols. One horizontal and one vertical line show the location of the splits for the input variables. Fig. 8.4 (top panel) shows a partition plot using the PEF and NPHI well logs corresponding to the tree shown in Fig. 8.2. The classification is refined further by incorporating DT as shown in the 3D partition plot (Fig. 8.3, bottom panel), which suggests that a majority of the electrofacies can be identified using just three well logs, that is, PEF, NPHI, and DT. Note that the classification problem is somewhat challenging because of the overlapping nature of the electrofacies clusters.

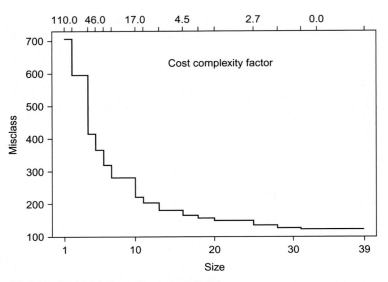

FIG. 8.3   **Example of pruning chart.** *From Perez et al. (2005).*

As an example of a *regression tree*, consider the two-variable surface generated using the function $\{y = \sin(x_1)^*\cos(x_2) + x_2^2/(4\pi) - x_1^2/(3\pi)\}$ and the corresponding optimal tree based on a random sample of 100 points drawn from this surface (Fig. 8.5) (SAMPLE_FIG8-5.DAT). As expected, the regression tree has a "blocky" nature, since it can only represent the continuous space of the response variable using a set of discrete values (in this case, corresponding to 58 terminal nodes). Regression trees are therefore best used as a tool for high-level understanding and as a building block for more advanced tree-based ensemble modeling approaches such as random forests or gradient boosting machines to be discussed next.

8.2.2 Random Forest

Random forest (RF) regression generates an ensemble of trees to increase performance of a single regression tree using a "bagging" (*bootstrap agg*regation) approach (Breiman, 2001a). Since using the entire input dataset would always yield the same regression tree, variation is introduced by using subsets of the input data and/or predictors to build multiple trees and thus view the dataset from these multiple perspectives as an ensemble or a "random forest." In practice, each tree in the ensemble is trained using a bootstrap sample of the training data, and a random subset of the predictors is considered for each split. This randomization allows each regression tree to focus on subtly different aspects of the predictor-response relationship. In aggregate, the trees can combine this information into a powerful prediction tool via an averaging step that reduces the variance from the noisy nature of individual trees.

The starting point for building an RF regression model is a set of regression trees, each of which is created from random subsets of data points and predictors, using the regression tree-building methodology described in the previous section. For prediction, each new

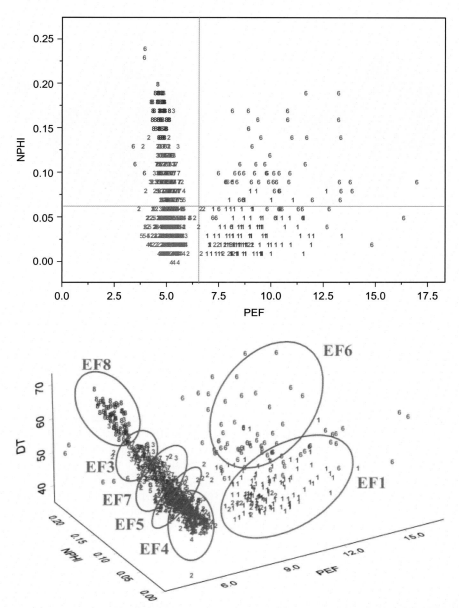

FIG. 8.4 **Example of two-dimensional (PEF-NPHI) and three-dimensional (PEF-NPHI-DT) partition plots.**
From Perez et al. (2005).

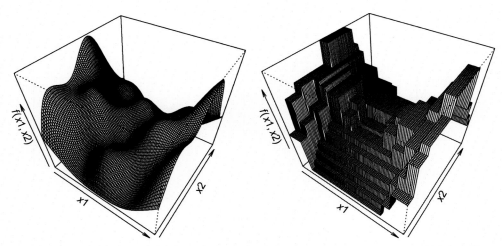

FIG. 8.5   **Example of regression tree, pruned to 58 terminal nodes (right), based on a random sampling of 100 points taken from a 2D surface (left).**

observation is passed through all the trees in the ensemble, thus producing a different regression estimate. The final model prediction is an average of those individual tree-level estimates. The predictive model is readily validated using the built-in cross validation capability in the RF algorithm (see Section 8.3.2 for an example of the cross validation procedure). Since each tree sees only a subset of the data, the remaining observations are called *out-of-bag samples*. For that tree, those out-of-bag samples can be treated as independent test data and used to develop estimates of error rates to gauge model performance.

Assignment of missing values for the predictors (also referred to as imputation) is handled using the concept of proximity. The proximity statistic used in the RF algorithm is a measure of similarity of different data points to one another represented via the normalized Euclidean distance in the form of a symmetrical matrix with 1 on the diagonal and values between 0 and 1 off the diagonal. The imputed value is the average of the nonmissing observations weighted by the corresponding proximities. Other than the imputation, the setup of RF is quite user-friendly and only involves two parameters: (1) number of predictor variables in the random subset at each node and (2) total number of trees in the ensemble.

The RF classifier is trained the same way as it was in the regression setting, with the exception that classification trees are used instead of regression trees since the response variable is categorical in nature. As shown in Fig. 8.6, classification trees are built using the approach described in the previous section. For prediction purposes, each observation is first passed through all of the trees in the ensemble, with each tree producing a predicted class label. The final label is the most popular vote among the trees. As before, out-of-bag samples are used to estimate misclassification rate on independent test data as part of cross validation. Additional computational details regarding the construction of the RF model and its interpretation can be found in Hastie et al. (2008).

An example of RF model, corresponding to the regression tree shown earlier in Fig. 8.5, will be discussed in the next section.

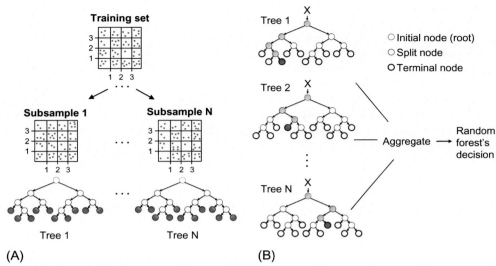

FIG. 8.6   **Schematic of random forest model-building procedure.**

8.2.3 Gradient Boosting Machine

Gradient boosting machine (GBM) for regression are similar to RF models in the sense that they are also ensembles of regression trees (Friedman, 2001). The basic idea in GBM is to gain prediction power from a large collection of simple models instead of building a single complex model. However, these trees are constructed sequentially, rather than in parallel as in the case of the RF model. Each new tree is constructed in such a way as to compensate for the shortcomings of the previous tree. In other words, when one tree tends to fit poorly to the training data for certain predictor values, the next tree will put more emphasis on observations in that problem area and ensure that the predictions have better accuracy. The final model can be considered a linear regression model with thousands of terms, where each term is a regression tree. This process is generally referred to as "boosting," wherein the outputs of many weak models are combined to produce a more accurate "committee" or aggregated prediction (Hastie et al., 2008).

The general GBM procedure involves starting with a base model (i.e., tree) and introducing a correction term (i.e., new model) to compensate for the residuals of the previous tree as identified by negative gradients of a squared-error loss function. The sequential fitting process can be repeated multiple times, with the caveat that the GBMs will soon start to model the noise and overfit. This problem can be handled in a variety of ways, namely, (a) using a fractional multiplier or learning rate on the correction term so the updated model improves the fit at a slower pace, (b) imposing constraints on the fitting parameters such as the maximum number of iterations, and (c) using a bootstrap sample of the data at each iteration rather than the full dataset.

The missing value issue is handled in GBMs with the construction of surrogate splits. The key step in tree-based modeling is to choose which predictor (and where) to split on at each node. In the GBM algorithm, nonmissing observations are used to identify the primary split, including the predictor and the optimal split point, and then form a list of surrogate splits that tries to mimic the action of the primary split. Surrogate splits are stored with the nodes and serve as a backup plan in the event of missing data. If the primary splitting predictor was missing during modeling or prediction, the surrogate splits will be used in order. The surrogate split utilizes the correlation between predictor variables to mitigate the negative impact of missing values.

GBMs are easily ported over from regression to the classification setting. The basic building block of the GBM classifier is a classification tree. Rather than fitting a single model at each step, there are multiple trees, one for each group. The negative gradient required for model updates is based on multinomial deviance rather than regression residuals. Additional computational details regarding the construction and interpretation of GBM models can be found in Hastie et al. (2008).

Examples of RF and GBM models that attempt to fit the surface shown in Fig. 8.5 are shown below (Fig. 8.7). The RF model uses 500 trees, whereas the GBM is based on 150 trees. As a

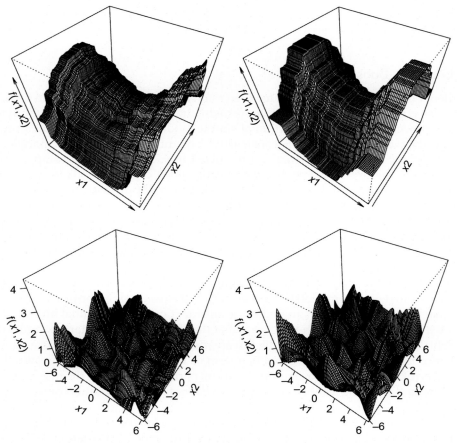

FIG. 8.7 **RF (left) and GBM (right) model predictions for the 2D surface shown in Fig. 7.5 (top panel) and corresponding residuals (bottom panel).**

result, the RF model predicted surface is somewhat smoother than the GBM version (top panel), even though the error rates are quite similar (bottom panel).

8.2.4 Support Vector Machine

Support vector machines (SVMs) are powerful machine learning tools for data classification and prediction (Vapnik, 1995). The problem of separating two classes is handled using a hyperplane that maximizes the margin between the classes (Fig. 8.8). The data points that lie on the margins are called support vectors. The SVM algorithm seeks to find the hyperplane that creates the biggest margin between the training points for the two classes. It also penalizes the total distance of points on the wrong side of their margin whenever there is overlap among the two classes of data. This permits a limited number of misclassifications to be tolerated near the margin.

The other key computational feature in SVM is the use of kernel functions and penalty parameter to convert nonlinear boundaries in the parameter space of the inputs to linear boundaries in some higher-dimensional transformed space. A popular choice in SVM applications is the radial basis function, which is described in Chapter 7 in the context of response surface modeling.

Fig. 8.9 illustrates the representation of a two-class problem in two-dimensional space using SVM (SAMPLE_FIG2-9.DAT). Here, the demarcation of boundaries between the red and blue classes (left panel) shows a predominantly continuous space for the red class with embedded blue pockets. The fitted SVM model (right panel) also creates a diagonally dominant pattern, albeit one where the blue class is continuous. The relative fraction of blue versus red space is very similar in both cases.

The concept of support vector regression (SVR) is very similar to that of SVMs. SVR machines are linear models where the parameters are optimized with respect to ε-insensitive loss, which considers any prediction within ε of the true value to be a perfect prediction (i.e., zero loss). During parameter estimation, the support vectors are also selected from the training dataset. Since the model is only specified through a dot product of support vectors and predictors, the "kernel trick" can also be used to transform the data to a linearized space, thus enabling highly nonlinear regression fits to be produced in the original input

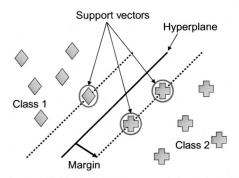

FIG. 8.8 Schematic showing separation of two classes of data using an optimal hyperplane in SVM.

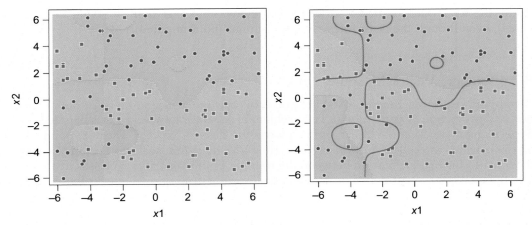

FIG. 8.9   **Test data for an example of two-class separation problem (left) and corresponding SVM model fit using a radial basis kernel function (right).**

feature space. Additional details regarding the computation of SVMs for regression and classification can be found in Hastie et al. (2008).

8.2.5 Artificial Neural Network

Artificial neural network (ANN) is a popular machine learning algorithm that attempts to mimic how the human brain processes information (Rumelhart and McClelland, 1986). It provides a flexible way to handle regression and classification problems without the need to explicitly specify any relationships between the input and output variables. Generally, neural networks are arranged in three layers: one input layer, one or more hidden layers, and one output layer—as shown in Fig. 8.10. In this example, the inputs are estimated attributes from various well logs, and the output is an indicator referring to the specific lithofacies assigned to that depth. For a regression problem, the output could be a numerical value (e.g., corresponding log- or core-derived permeability).

In the ANN, each layer contains a number of nodes (or artificial neurons) that are connected to each of the nodes in the preceding layer by simple weighted links. Except for nodes in the input layer, each node multiplies its specific input value by the corresponding weight and then sums all the weighted inputs. Sometimes, a constant (the "bias" term) can be involved in the summation. The final output from the node is calculated by applying an activation function (transfer function) to the sum of the weighted inputs.

A critical aspect in neural network modeling is the learning process of forcing a network to yield a particular output (response) for a specific input (signal). ANN modeling starts with randomly assigned weight coefficients. Then, a set of data patterns are fed forward repeatedly, and the weights of the neurons are modified until the output matches closely with the actual values. For multilayer feed-forward neural networks, a more powerful supervised learning algorithm, called backpropagation, can be employed to recursively adjust the connection weights so that the difference between the predicted and the observed outputs

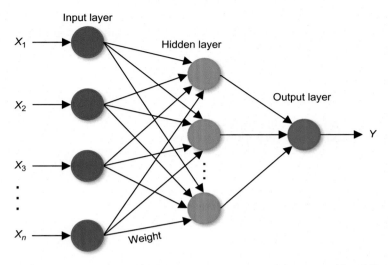

FIG. 8.10   Schematic architecture of an ANN algorithm where input layers correspond to different well logs and output refers to different lithofacies.

is as small as possible. The most important parameters to control in building the ANN include the number of hidden layers, number of hidden layer nodes, learning rate, damping coefficient or momentum, and number of iterations for better optimization. See Hastie et al. (2008) for additional computational details regarding the construction of ANNs for both classification and regression problems.

We revisit the classification problem example discussed earlier, to show the performance of two different ANNs in Fig. 8.11. The left panel shows a simple ANN with one hidden layer and two hidden units, which fails to capture the actual class boundaries as indicated by the

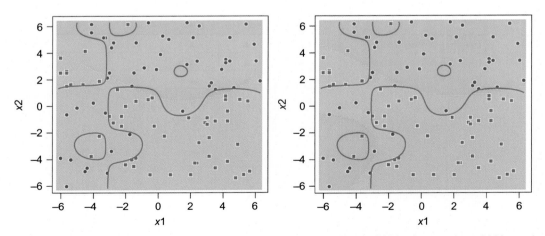

FIG. 8.11   ANNs for the example of two-class separation problem with one hidden layer and two hidden units (left) and one hidden layer and five hidden units (right).

green line. A more complicated ANN with one hidden layer and five hidden units (right panel) is much better in preserving the structure of the data, and its performance is very similar to that of the SVM model shown in Fig. 8.9.

8.2.6 Model Strengths and Weaknesses

The five data-driven modeling approaches discussed in this chapter, that is, CART, RF, GBM, SVM, and ANN, are very powerful tools for both regression and classification problems. However, there are important differences with respect to key performance attributes such as the following:

- Ability to handle missing data and missing values
- Robustness to outlier data points and irrelevant inputs
- Insensitivity to monotone transformation of inputs
- Ability to extract linear combinations of features
- Computational scalability
- Interpretability
- Predictive power

Table 8.1 provides a compact summary of how the five different modeling approaches stack up against these attributes, using a format originally presented by Hastie et al. (2008). Broadly, it can be concluded that CART can be useful for preliminary modeling and only as basic model building blocks—primarily because of their poor predictive power. On the other hand, the two ensemble tree-based methods RF and GBM have good predictive power and a number of desirable features related to computational robustness (e.g., handling of missing data and robustness to outliers). While SVM and ANN also have good predictive

TABLE 8.1   **Comparison of Model Strengths and Weaknesses**

	CART	RF	GBM	SVM	ANN
Handling of mixed data	▲	▲	▲	▼	▼
Handling of missingvalues	▲	▲	▲	▼	▼
Robustness to outliers	▲	▲	▲	▼	▼
Insensitivity to monotone transformations of inputs	▲	▲	▲	▼	▼
Computational scalability	▲	♦	♦	▼	▼
Ability to deal with irrelevant inputs	▲	▲	▲	▼	▼
Ability to extract linear combinations of features	▼	▼	▼	▲	▲
Interpretability	♦	▼	▼	▼	▼
Predictive power	▼	▲	▲	▲	▲

▲ ,good; ♦ ,fair; ▼ ,poor

power, they do not offer the same degree of computational robustness as RF and GBM. Unfortunately, all four approaches (RF, GBM, SVM, and ANN) suffer from the problem of poor interpretability.

How then to balance the conflict between predictive power and interpretability? Since these models are essentially black-box solutions, one approach would be to develop greater insights into each model's internal architecture. Some useful strategies to this end are (a) determining the relative importance of each predictor with respect to the response of interest (see Section 8.3.3) and (b) performing a conditional sensitivity analysis to better understand how the variation of any given predictor affects the model response when other correlated inputs are varied commensurately while uncorrelated input parameters are held at their mean/median values (see Section 8.4.4).

8.3 COMPUTATIONAL CONSIDERATIONS

8.3.1 Model Evaluation

A common approach to evaluating the goodness of fit for a model is to generate a scatterplot of actual response values in the training dataset versus the corresponding predicted (fitted) responses. If the points in the scatterplot lie close to the 45-degree (1:1) line, this indicates a good model fit to the training data. However, this does not necessarily indicate that the model has good predictive ability for new datasets. Consider the model shown in Fig. 8.12 (left), which is clearly overfitting the training dataset by trying to capture not only the true underlying function but also the noise in the measurements. The model likely contains more degrees of freedom than are necessary to capture the underlying shape of the curve producing these observations, which makes it unlikely to have good predictive power going forward. However, in the model evaluation scatterplot (shown in the right panel), all the points lie along the 45-degree line, indicating a very good fit. The point of this

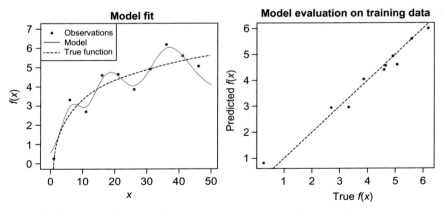

FIG. 8.12   **Example of a poor model that appears to fit well when evaluated solely against the training set but could be suffering from overfitting (Schuetter et al., 2015).**

simple example is to emphasize the risks of overfitting and the need to move beyond using predictions on the training data itself as the sole measure of model quality.

One simple strategy for accomplishing this is to evaluate the model using an independent test set. This can be either a completely new dataset (e.g., pilot data from a region where the model is intended for use) or a "held out" portion of the training dataset. In both cases, one can fit the model using the training portion of the dataset (typically 70%–90% of the data) and then evaluate the fit on the independent test observations (i.e., remaining 10%–30% of the data) to gauge the predictive ability of the model for new data. The challenge is in ensuring that the test set is sufficiently broad enough to cover the full range of potential applications of the model.

A better choice for model evaluation is *k*-fold cross validation (Hastie et al., 2008). In this approach, schematically shown in Fig. 8.13, the training dataset is randomly split into *k* different groups or "folds." Next, each of the *k* groups is held out one at a time; the model is trained on the remaining *k* − 1 groups and used to make predictions on the group that was held out. After cycling through all *k* groups, there will be a single cross validated prediction for every observation in the dataset, where the predictions were made using a model for which that observation was not included in the training set.

It is important to note that the cross validation procedure can be extended by repeating the entire process with a different random selection of *k* groups. A repeated cross validation using *r* repeated runs of *k* randomly selected groups will yield *r* different predictions on each of the observations. Not only these can be aggregated to compute statistics on goodness-of-fit metrics, but also they give important insight into the variability in model predictions depending on the characteristics of the training set. Also, the models trained during cross validation are not the models to be used for prediction going forward; rather, one would build a single predictive model using the full training set. The cross validation procedure is only for evaluation purposes and provides a better indication of the robustness of the predictive model for future applications using new data, as compared with a single heldout test set.

Next, we discuss three common metrics for quantifying the goodness of fit: (1) average absolute error or AAE, (2) mean squared error or MSE, and (3) pseudo-R^2. These metrics are broadly similar, in that they attempt to capture the overall closeness of predictions to the evaluation data. Let y_i be the true response for the ith observation and $\hat{y}_i$ be the predicted response

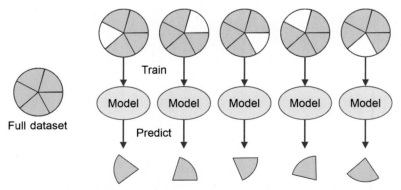

FIG. 8.13   **Conceptual representation of *k*-fold cross validation with *k* = 5 (Schuetter et al., 2015).**

for that observation. The AAE is defined as the average magnitude of the difference between the true response and predicted response (i.e., the average size of the residuals), as in Eq. (8.1):

$$AAE = \frac{1}{n} \sum_{i=1}^{n} |y_i - \hat{y}_i| \tag{8.1}$$

MSE is similar to AAE but measures the average squared difference between observations and their corresponding predictions, rather than the absolute value:

$$MSE = \frac{1}{n} \sum_{i=1}^{n} (y_i - \hat{y}_i)^2 \tag{8.2}$$

Note that AAE has units matching those of the response, while MSE is measured in squared units of the response. A common variant of MSE is the root-mean-square error or RMSE, which is simply the square root of MSE. Values closer to zero are desirable, as they indicate smaller deviations between the observations and predictions (i.e., more accurate prediction). MSE (or RMSE) is typically preferred over AAE due to its well-known distributional properties and being a sufficient statistic for normally distributed random processes (Navidi, 2008).

The third metric, pseudo-R^2, is defined in Eq. (8.3):

$$R_p^2 = 1 - \frac{SS_{model}}{SS_{total}} = 1 - \frac{\sum_{i=1}^{n} (y_i - \hat{y}_i)^2}{\sum_{i=1}^{n} (y_i - \bar{y})^2} \tag{8.3}$$

Pseudo-R^2 compares the sum of squared differences between the true responses y_i and predicted responses $\hat{y}_i$ with the overall sum of squares, which is proportional to the variance of the responses. That is, it measures how much of the variability in the response is explained by the model. Note that while in linear regression the pseudo-R^2 is bounded between 0 and 1, this is not the case for a general regression model. When a regression model fits the data worse than a constant value at the mean response, the pseudo-R^2 will be negative.

8.3.2 Automatic Tuning of Model Parameters

Selecting the values of the tuning parameters in various "black-box" data-driven modeling algorithms often becomes a manual time sink, with the added potential for significant subjective bias. Examples of such tuning parameters include (a) number of variables randomly sampled as candidates at each split and number of trees for the RF algorithm, (b) number of trees for the GBM algorithm, (c) cost parameter for the SVM algorithm, and (d) number of hidden layers and hidden units for the ANN algorithm.

To his end, an automated process that relies on cross validation has been suggested by Kuhn and Johnson (2013). The basic steps are as follows: (a) define a set of candidate values for the tuning parameters(s); (b) for each candidate set, resample data, fit model, and predict hold outs for a k-fold cross validation strategy; (c) aggregate the resampling into a performance profile; (d) determine the final tuning parameters using cross validated RMSE as the accuracy metric of choice; and (e) using the final tuning parameters, refit the model

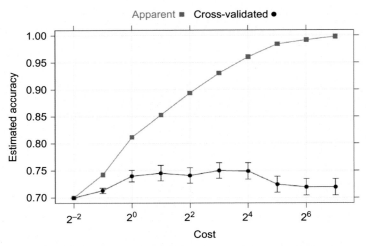

FIG. 8.14 **Example showing automatic tuning of SVM model parameters using normalized cross validated RMSE as the accuracy metric.**

with the entire training set. Fig. 8.14 shows an example application of this strategy for an SVM model for selecting the optimal value for the cost parameter.

8.3.3 Variable Importance

When data-driven models are fitted to large multivariate datasets, the resulting interactions of variables with one another can be complex and/or nonlinear. As such, it is difficult to develop a straightforward understanding of causal input-output relationships and key sensitivities based on a simple evaluation of model results. The problem is also compounded by the fact that often only a few predictors have any significant influence on the model response, making the others largely irrelevant. It is therefore useful to have a strategy that complements the building of a predictive model for a given response by determining the relative importance of each input among a set of predictors. This helps the analyst focus on the key variables for future data collection efforts and screen out unimportant variables during subsequent iterations of the model-building process.

In general, the identification of variable importance tends to be model-specific, and the corresponding metrics can be specified in absolute or relative units. For example, the relative importance computed for an RF model measures the prediction strength of each variable by calculating the increase of RMSE when that variable is permuted while all others are left unchanged (Breiman, 2001a). The rationale behind the permutation step is that if the predictor variable was not important to the tree-building process, rearranging the values of this variable will not change the prediction accuracy much. On the other hand, relative importance for GBM models is based on the number of times a predictor variable was selected for splitting, weighted by the squared improvement to the model as a result of each split, averaged over all trees, and rescaled with a total sum of 100 (Friedman, 2001).

One straightforward approach for variable importance that is not tied to any particular model is based on the concept of R^2-loss (Mishra et al., 2009). This method works for any regression model, and the reasoning is that if an influential predictor is removed from a model, the accuracy of that model will be significantly reduced. Alternatively, if a superfluous predictor is removed from the model, there should be little to no impact on the accuracy. To measure variable importance, one can compute R_p^2 (i.e., pseudo-R^2 as defined earlier in Eq. 8.3) using all the predictors and then compute R_p^2 for a reduced model that uses all of the predictors *except* the predictor of interest. The "R^2-loss" metric is simply the difference between R_p^2 for the full model and R_p^2 for the reduced model. The larger the loss in pseudo-R^2 for any given predictor, the greater its influence on the model response.

8.3.4 Model Aggregation

In our experience with data-driven model applications, it is quite common to have several modeling techniques resulting in comparable goodness of fit in terms of AAE, RMSE, or pseudo-R^2. Since this does not necessarily imply comparable performance for a new set of data, the question is to how to pick one model from the collection for future applications. As an alternative, one could consider the possibility of aggregating the predictions over a number of models using some statistical model averaging procedure. Such approaches are increasingly being applied in the context of subsurface flow and transport modeling to combine predictions from models that represent varying degrees of conceptual (geologic) model uncertainty (see Singh et al., 2010, and references therein). The goal of such an exercise is to determine the weighting for each model based on model performance against observed data and then develop an "ensemble" prediction by creating a weighted average of all the model predictions.

The problem of model averaging is generally handled using a Bayesian formalism (e.g., Draper, 1995), where the model weights, w_j, are given by

$$w_j = \frac{L_j \, p(M_j)}{\sum_j L_j \, p(M_j)} \tag{8.4}$$

Here, L represents the likelihood of the model that depends on the prediction error of the model for the given data, and $p(M_j)$ represents the prior probability of the model. In the case of data-driven modeling, all models can be assigned the same prior probability (i.e., $p(M_j) = 1/N$) where N is the total number of models under consideration.

One formal approach to the determination of model likelihoods is via maximum likelihood Bayesian model averaging or MLBMA (Neuman, 2003). The starting point for MLBMA is a collection of models that have been calibrated to observed data using maximum likelihood estimation. The likelihood for each model is then estimated using

$$L_j \propto \exp\left(-\frac{\mathrm{BIC}_j - \mathrm{BIC}_{\min}}{2}\right) \tag{8.5}$$

where the difference is taken between the Bayesian information criterion (BIC) measure for the jth model and the minimum BIC value among all competing models. Assuming a multi-Gaussian error distribution with unknown mean and variance for the model likelihood, the BIC term for model j can be written as

$$\mathrm{BIC}_j = (n) \, \ln\left(\hat{\sigma}^2_{e,j}\right) + k_j \, \ln(n) \tag{8.6}$$

with n, number of observations; k, number of model parameters; and σ_e^2, variance of residuals. Note that because of the exponential weighting in Eq. (8.5), the Bayesian model averaging approach tends to concentrate model weights on only one to two best-performing models.

A more practical alternative to model aggregation is the generalized likelihood uncertainty estimation (GLUE) procedure. It is based on the concept of "equifinality," that is, the possibility that the same final state may be obtained from a variety of initial states (Beven and Binley, 1992). In other words, a single set of observed data may be (nonuniquely) matched by multiple parameter sets that produce similar model predictions. Here, the likelihood for each model is also computed as a function of the misfit between observations and model predictions.

One of the central features of GLUE is the flexibility with respect to the choice of the likelihood measure. For example, two common choices are

$$L_j \propto \exp\left[-N\frac{\sigma_{e,j}^2}{\sigma_o^2}\right] \quad \text{or} \quad L_j \propto \left(\frac{\sigma_o^2}{\sigma_{e,j}^2}\right)^N \tag{8.7}$$

where L_j is the likelihood for model j, $\sigma_{e,j}^2$ is the variance of the errors (residuals) for model j, σ_o^2 is the variance of the observations, and N is a shape factor such that values of $N \gg 1$ tend to give higher weights (likelihoods) to models with better agreement with the data and values of $N \ll 1$ tend to make all models equally likely. A simpler version of Eq. (8.7) can be defined using the traditional root-mean-square-error (RMSE) statistic as follows:

$$L_j \propto \left(\frac{1}{\text{RMSE}}\right)^2 \tag{8.8}$$

The aggregated model response can then be calculated as a weighted average of the responses from multiple models, using the likelihood relationships given in Eq. (8.7) or (8.8) (see Mishra, 2012, for an application in decline curve analysis).

8.4 FIELD EXAMPLE

8.4.1 Dataset Description

The techniques described in this section will be illustrated on an example dataset from West Texas, the United States (Zhong et al., 2015; Schuetter et al., 2015). The study area is the Delaware Basin, where the Wolfcamp shale forms an unconventional reservoir of roughly 2000–4000 ft thick, which is being exploited by a number of horizontal wells.

A publicly available dataset of 476 horizontal shale wells from phantom field was selected for this study. The predictor variables relate to operational characteristics of the wells, including when the well was drilled, its physical dimensions, stimulation details, and by whom it is operated. The response measures cumulative well production (in barrels) over the first 12 producing months. A list of all variables in the dataset is shown in Table 8.2.

Next, we discuss the process of building predictive models for M12CO as a function of the predictors listed in Table 8.2. This is followed by a classification tree analysis to identify the

TABLE 8.2   List of Variables in the Study Dataset

Type	Variable	Description
–	ID	Well identification number
Response	M12CO	Cumulative production within first 12 producing months (BBL)
Predictor	Opt2	Categorized operator code
	COMPYR	Well completion year
	SurfX	Geographic location (horizontal)
	SurfY	Geographic location (vertical)
	AZM	Azimuth angle (degrees)
	TVDSS	True vertical depth (ft)
	DA	Drift angle (degrees)
	LATLEN	Total horizontal lateral length (ft)
	STAGE	Number of frac stages
	FLUID	Total frac fluid amount (gal)
	PROP	Total proppant amount (lb)
	PROPCON	Proppant concentration (lb/gal)

key attributes that separate good wells (i.e., those corresponding to the top 25% of M12CO values) from bad wells (i.e., bottom 25% in terms of M12CO).

8.4.2 Predictive Model Building

Before starting the model-building process, missing values for the predictors found in 157 of the wells were imputed to create a complete dataset. Predictive models were built using three data-driven algorithms, that is, RF, GBM, and SVR, a multilinear regression model described in Chapter 4 (referred to here as ordinary least squares or OLS) and a multidimensional kriging model described in Chapter 7 (referred to here as kriging meta model or KM). Model-fitting results are summarized in Fig. 8.15. Each plot shows the true response (M12CO) on the horizontal axis and the predicted response on the vertical axis. Points on the diagonal dotted line indicate perfect prediction. Each row of plots shows predictions from one type of model (OLS, RF, GBM, SVR, and KM), while each column shows results for a different model evaluation type.

The left column shows independent validation results, where a random 20% subset of the wells was held out to create a single heldout test data. The model was then fit to the remaining 80% of the dataset and evaluated on the 20% heldout set. The points in the plots in the left column only correspond to those predictions on the holdout segment of the dataset. For the cross validation predictions (center column) a 10-fold cross validation was used as a further refinement of the fivefold cross validation approach discussed earlier. The points in these plots show the actual versus the cross validated predictions of each of the wells in the dataset. The right column shows the results from training and predicting on the full dataset, which is the conventional approach to evaluating goodness of fit.

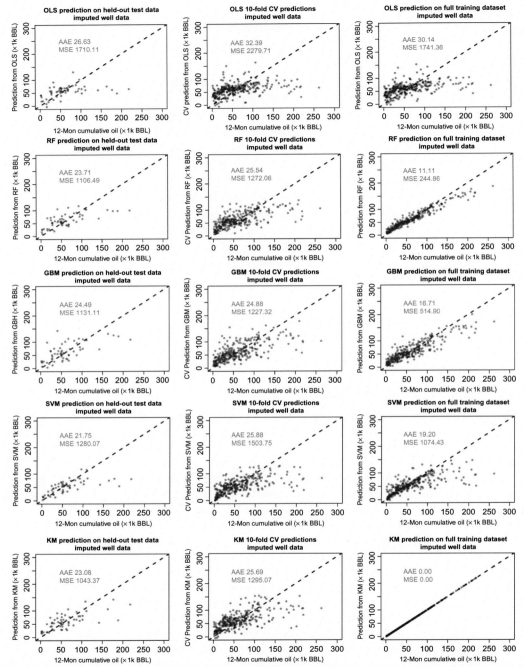

FIG. 8.15 **Comparison of model performance for OLS, RF, GBM, SVR, and KM models using different model evaluation approaches.** *After Schuetter, J., Mishra, S., Zhong, M., LaFollette, R., 2015. Data analytics for production optimization in unconventional reservoirs. In: Proc. SPE/AAPG/SEG Unconventional Resources Technology Conference, San Antonio, TX, July 21–23.*

Notice that the independent validation and cross validation results (left and center columns) tell a much different story than the predictions on the full training set (right column). Except for OLS, all other models show a dramatic reduction in error in both the AAE and MSE metrics—if one accepts the full training set approach for model evaluation. However, this reduction is more modest for the other methods of model evaluation. The extreme case is the kriging model (KM), which is a perfect interpolator and hence, by design, forces the model fit through the training observations. However, for a case like the random forest (RF), it is not so clear that the predictions on the training data are biased. It is only when comparing the goodness of fit for the full training set to those shown in the independent validation and/or cross validation plots that the overfitting is revealed. As noted earlier, the fit statistic for a multifold cross validation is more likely to be a robust indicator of predictive model performance on new datasets.

8.4.3 Variable Importance and Conditional Sensitivity

We now discuss the relative importance of different predictors using the R^2-loss metric. When determining variable importance, it can be useful to compute the ranks using several different predictive models to get a more robust sense of which predictors are important. In this case, there is some divergence among the four selected models (i.e., OLS, RF, GBM, and SVR) as to which predictors are the most influential. The depth parameter (TVDSS) is popular among all models. Three of the four models also put weight on the amount of proppant used (PROP), the length of the lateral (LATLEN), and the amount of fracturing fluid used (FLUID). A compact way of visualizing this information about variability of ranking across models is through horizontal box plots, as shown in Fig. 8.16. The box plots are sorted from bottom to top by average rank, with the width displaying the degree of variation. TVDSS is clearly an influential predictor, with high rank and low variability. FLUID also has a reliable rank in the middle of the pack. Finally, opt2A is definitely not important, with consistently low rankings.

The variable importance analysis reflects model performance across the entire range of predictor and response variables, converted into a set of ordinal ranks. However, model interpretability can be further enhanced by explaining how the model response changes as the value of any of the predictor variables is changed. This is similar to a standard one-parameter at a time (OPAT) sensitivity analysis. A more meaningful strategy is to perform a conditional sensitivity analysis to quantify the model response for a specified variation of any given predictor, when other correlated inputs are varied commensurately while the uncorrelated input parameters are fixed at some reference values. For the Wolfcamp dataset, this involves varying the key parameters COMPYR, LATLEN, FLUID, and PROP in a correlated manner (reflecting the relationship in the observed data) while setting the other variables at the mean or median values. Fig. 8.17 shows these conditional sensitivity analysis results for M12CO generated using the SVM model (blue line), whereas the standard OPAT analysis results are shown in the red line. The background symbols are those for the original dataset. It is clear that the conditional sensitivity analysis is a more robust approach to answering "what-if" questions as it forces the analyst to vary predictors whose correlations cannot be ignored (e.g., LATLEN and PROP/FLUID).

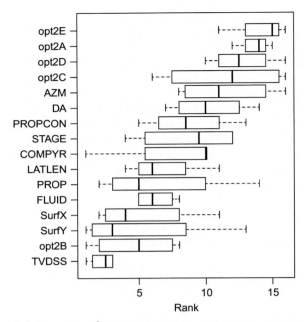

FIG. 8.16   **Importance ranking based on R^2-loss metric and aggregated across OLS, RF, GBM, and SVR models.**
*After Schuetter, J., Mishra, S., Zhong, M., LaFollette, R., 2015. Data analytics for production optimization in uncon-
ventional reservoirs. In: Proc. SPE/AAPG/SEG Unconventional Resources Technology Conference, San Antonio, TX,
July 21–23.*

8.4.4 Classification Tree Analysis

Having dealt with building a predictive regression model for cumulative first year production (M12CO) in terms of a given set of well characteristics, we now turn to the issue of trying
to predict whether the well will performance be a "good" (i.e., relatively large M12CO) or
"bad" (i.e., relatively low M12CO). This can be accomplished by changing the problem from
regression to classification. That is, the response can be binned into categories, and classifier
models (e.g., classification tree) can be used to predict which category a well falls into.

For the Wolfcamp dataset, the top 25% and bottom 25% producing wells were identified,
and the middle 50% of the wells were removed. A classification tree was then built to separate
the top and bottom 25% groups, with the result shown in Fig. 8.18. The tree begins at the top of
the figure, where the first split checks whether the proppant used is less than 1.405e6 lb. If so,
a well observation moves down the left path; otherwise, it goes right. Subsequent splits work
in the same way, until eventually the observation reaches a terminal node that contains a prediction. The text at the terminal nodes in this tree indicate how many training observations on
each type (bottom 25%/top 25%) ended up in that node.

One advantage of classification trees (compared with other common classifiers such as
SVM and ANN) is better interpretability. Not only do they clearly indicate which predictors
are influential in determining the response category, but also they identify critical values at
which these categories change. As shown in Fig. 8.18, there are two general paths to a top 25%
producing well. For wells using lower amounts of proppant (PROP < 1.405e6 lb), the goal is

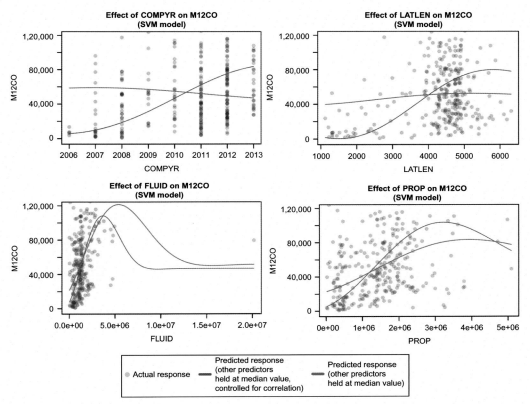

FIG. 8.17   Comparison of model performance for OLS, RF, GBM, SVR, and KM models using different model evaluation.

to have a longer lateral (LATLEN ≥ 2756 ft) and a greater vertical depth (TVDSS < −8294 ft). For wells using more proppant (PROP ≥ 1.405e6 lb), the goal is again to have a greater vertical depth (TVDSS < −8100 ft) and to have a lateral that is not too long (LATLEN < 5362 ft).

Fig. 8.19 shows partition plots that render the classification tree from the perspective of the wells in the predictor space. In these scatterplots of two predictors, the predictor space is partitioned into blocks of similar observations via vertical and horizontal segmentations of the plot. For example, in the top left plot, the first split at PROP = 1.405e6 appears as a vertical division of the plot. Within each of those divisions, the splits on LATLEN (2756 on the left branch and 5362 on the right) serve to further subdivide the plot into relatively homogeneous data clusters. In general, the tree is fairly efficient at partitioning the predictor space into the region that contains primarily the top 25% wells.

A "confusion matrix" that summarizes the separability of the two classes in the training set is shown in Table 8.3. The value in each cell describes how many wells of the true category indicated in the row header were in a terminal node whose majority category was the one indicated in the column header. Since 62 of the 80 true top 25% wells were in "top 25%" terminal nodes, this yields a correct identification rate of 62/80 = 77.5%. A similar calculation

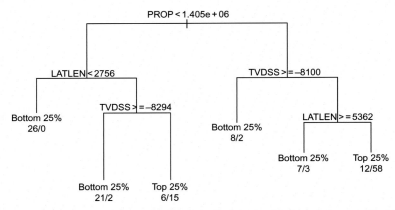

FIG. 8.18 **Classification tree separating top 25% of wells for M12CO from bottom 25%.** *After Schuetter, J., Mishra, S., Zhong, M., LaFollette, R., 2015. Data analytics for production optimization in unconventional reservoirs. In: Proc. SPE/AAPG/SEG Unconventional Resources Technology Conference, San Antonio, TX, July 21–23.*

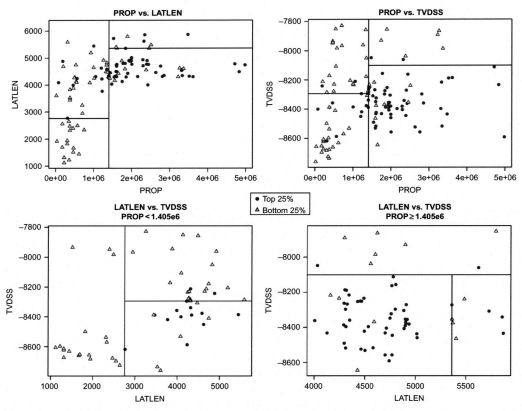

FIG. 8.19 **Partition plots showing the separation of top 25% wells** *(circle)* **and bottom 25% wells** *(triangle)* **in the 2D space for PROP-LATLEN, PROP-TVDSS, and LATLEN-TVDSS pairs.** *After Schuetter, J., Mishra, S., Zhong, M., LaFollette, R., 2015. Data analytics for production optimization in unconventional reservoirs. In: Proc. SPE/AAPG/ SEG Unconventional Resources Technology Conference, San Antonio, TX, July 21–23.*

TABLE 8.3   Classification Tree Terminal Node Confusion Matrix

	Predicted Bottom 25% Wells	Predicted Top 25% Wells	Total	Correct ID Rate (%)
True bottom 25% wells	62	18	80	77.5
True top 25% wells	7	73	80	91.3
Total	69	91	160	84.4

gives a correct identification rate of 91.3% for the bottom 25% wells. Overall, the rate is $(62+73)/160=84.4\%$. This indicates a reasonable ability to separate the two classes.

8.5 SUMMARY

In this chapter, we have presented a number of commonly used data-driven modeling approaches for both regression and classification problems. These include classification and regression tree, random forest, gradient boosting machine, support vector machine, and artificial neural network. We also discussed several computational considerations such as model evaluation, automatic tuning of model parameters, identifying variable importance, and model aggregation. Finally, a field application was used to demonstrate the practical applicability of these algorithms for predictive modeling, variable importance, conditional sensitivity, and classification.

Exercises

Prerequisites: Download the folder named "Data_Driven_Modeling" from the online resources website for this book. As a prerequisite R needs to be installed on the user computer. *R Studio* should be installed in order to edit code if needed. Also, following list of libraries are needed to be installed: "xlsx," "Metrics," "randomForest," "e1071," "MASS," "gbm," "ggplot2," "cvTools," "class," "maps," "devtools," "rpart.plot," "reshape2," and "neuralnet."

In order to install a library, go to *R Studio* menu bar and press *Tools* → *Install Packages.* A window should be opened up where the needed library can be installed. In case a library is needed but not installed, *R Studio* should generate error in console.

For the problems below, use the data file "Model_data.xlsx." List of predictors to be used in the following problems are:

Predictor	Description
qi	Initial flow rate of the well (STB/month)
PROP_TOTAL	Total amount of proppant per well (lbs)
FRAC_FLUID_TOTAL	Total amount of fracturing fluid per well (bbl)

CLENGTH	Difference between measured depths of last and first perforation (ft)
STAGES	Number of hydraulic fracture stages (dimensionless)
TVD_HEEL	Total vertical depth of well heel (ft)
TVD_HEEL_TOE_DIFF	Difference between total vertical depths of well heel and well toe (ft)
LATITUDE	Latitude of well head location (degrees)
LONGITUDE	Longitude of well head (degrees)

1. Train a regression tree model to predict estimated ultimate recovery (EUR) calculated from stretched exponential decline model (SEDM) model (SEDM_EUR). Explain how you chose the cost-complexity parameter for this tree with illustration.
2. Cluster wells into four groups according to EUR calculated from SEDM model (SEDM_EUR). Train a classification tree model to predict cluster numbers (1, 2, 3, or 4) and explain the choice of cost-complexity parameter with illustration. Repeat using SVM and ANN. Comment on the relative performance of these methods.
3. Divide wells randomly into training data (80%) and test data (20%).

   a. Taking only the training data, build models using the machine learning algorithms: RF, SVM, GBM, and ANN to make predictions for "SEDM_EUR" and the decline curve parameters of SEDM: "tau" and "n."
   b. Make predictions for "SEDM_EUR," "τ," and "n" for training and test data wells and show the performance in a plot (i.e., actual vs. predicted). Also, report root mean squared error (RMSE) and pseudo R^2 error for train and test data fits. Examine the relative performance of the machine learning algorithms used. RMSE and pseudo R^2 are given by:

$$RMSE = \sqrt{\frac{1}{n} \sum_{i=1}^{n} (y_i - \hat{y}_i)^2}$$

$$\text{pseudo } R^2 = \frac{\sum_{i=1}^{n} (\hat{y}_i - \bar{y})^2}{\sum_{i=1}^{n} (y_i - \bar{y})^2}$$

where y_i = observed value of ith data point, $\hat{y}_i$ = predicted value of ith data point, and $\bar{y}$ = mean of observed values.

   c. Using the predicted values of "τ" and "n" for test data wells, plot actual well rates and predicted SEDM decline curves for each of the machine learning algorithms in a single time versus rate plot.

4. Comment on the relative performance of RF, SVM, GBM, and ANN as machine learning algorithms. Which machine learning algorithm gives the best performance? How does this change if you used a different 80–20 split to train and test the data?

5. Relative influence (RI) of pth predictor in a machine learning model is given by:

$$\mathrm{RI}_p = abs\left(\frac{R^2{}_p - R^2{}_{-p}}{R^2{}_p}\right)$$

where $R^2{}_p = \mathrm{pseudo}\,R^2$ of model with all predictors included and $R^2{}_{-p} = \mathrm{pseudo}\,R^2$ of model with all predictors except pth predictor are included.

Removing one predictor at a time, calculate RI measure of each of the predictors taking only test data R^2 into account. Rank the variables according to RI (e.g., predictor with highest RI is ranked 1 and so on). Repeat this for all four machine learning algorithms (RF, SVM, GBM, and ANN). Show the variation of predictors" RI using a boxplot and a histogram. Which predictors are making the highest influence?

References

Bahga, A., Madisetti, V., 2016. Big Data Science and Analytics: A Hands-On Approach. VPT. www.big-data-analytics-book.com.

Beven, K.J., Binley, A., 1992. The future of distributed models: model calibration and uncertainty prediction. Hydrol. Process. 6, 279–298.

Bhattacharya, S., Carr, T.R., Pal, M., 2016. Comparison of supervised and unsupervised approaches for mudstone lithofacies classification: case studies from the Bakken and Mahantango-Marcellus Shale. J. Natl. Gas Sci. Eng. 33. https://doi.org/10.1016/j.jngse.2016.04.055.

Breiman, L., 2001a. Random forests. Mach. Learn. 45 (1), 5–32.

Breiman, L., 2001b. Statistical modeling: the two cultures. Stat. Sci. 16 (3), 199–231.

Breiman, L., Friedman, J.H., Olshen, R.A., Stone, C.J., 1984. Classification and Regression Trees. Wadsworth and Brooks/Cole, Monterey, CA.

Draper, D., 1995. Assessment and propagation of model uncertainty. J. R. Stat. Soc. Ser. B 57 (1), 45–97.

Friedman, J.H., 2001. Greedy function approximation: a gradient boosting machine. Ann. Stat. 29, 1189–1232.

Hastie, T., Tibshirani, R., Friedman, J.H., 2008. The Elements of Statistical Learning: Data Mining, Inference, and Prediction. Springer, New York.

Holdaway, K., 2014. Harnessing Oil and Gas Big Data With Analytics. Wiley, Hoboken, NJ.

Kuhn, M., Johnson, K., 2013. Applied Predictive Modeling. Springer, New York.

Lolon, L., Hamidieh, K., Weijers, L., Mayerhofer, M., Melcher, H., Oduba, O., 2016. Evaluating the relationship between well parameters and production using multivariate statistical models: a Middle Bakken and Three Forks case history. Soc. Pet. Eng. https://doi.org/10.2118/179171-MS. SPE-179171-MS.

Maučec, M., Cullick, S., Shi, G., 2011. Geology-guided quantification of production-forecast uncertainty in dynamic model inversion. In: SPE Annual Technical Conference and Exhibition, Denver, Colorado, 30 October–2 November. https://doi.org/10.2118/146748-MS.

Maysami, M., Gaskari, R., Mohaghegh, S., 2013. Data driven analytics in Powder River Basin, WY. In: SPE Annual Technical Conference and Exhibition, New Orleans, LA, 30 September–2 October.

Mishra, S., 2012. A new approach to reserves estimation in shale gas reservoirs using multiple decline curve analysis models. In: SPE Eastern Regional Meeting, Lexington, KY, October 3–5.

Mishra, S., Deeds, N.E., Ruskauff, G.J., 2009. Global sensitivity analysis techniques for groundwater models. Ground Water 47 (5), 730–747.

Navidi, W., 2008. Statistics for Engineers and Scientists. McGraw Hill, New York.

Neuman, S.P., 2003. Maximum likelihood Bayesian averaging of uncertain model predictions. Stoch. Environ. Res. Risk A 17 (5), 291–305.

Perez, H.H., Datta-Gupta, A., Mishra, S., 2005. The role of electrofacies, lithofacies, and hydraulic flow units in permeability predictions from well logs: a comparative analysis using classification trees. Soc. Petrol. Eng. https://doi.org/10.2118/84301-PA.

Rawi, Z., 2010. Machinery predictive analytics. In: SPE Intelligent Energy Conference and Exhibition, Utrecht, The Netherlands, March 23–25.

Rumelhart, D.E., McClelland, J.L., 1986. Parallel Distributed Processing, 1: Foundations. MIT Press, Cambridge.

Santos, I.H.F., et al., 2015. Big data analytics for predictive maintenance modeling: challenges and opportunities. In: Offshore Technology Conference, Rio de Janiero, Brazil, October 27–29.

Saputelli, L., 2016. Technology focus: petroleum data analytics. Soc. Pet. Eng. https://doi.org/10.2118/1016-0066-JPT.

Schuetter, J., Mishra, S., Zhong, M., LaFollette, R., 2015. Data analytics for production optimization in unconventional reservoirs. In: Proc. SPE/AAPG/SEG Unconventional Resources Technology Conference, San Antonio, TX, July 21–23.

Shelley, R., Nejad, A., Guliyev, N., Raleigh, M., Matz, D., 2014. Understanding multi-fractured horizontal Marcellus completions. Soc. Pet. Eng. https://doi.org/10.2118/171003-MS.

Singh, A., Mishra, S., Ruskauff, G., 2010. Model averaging techniques for quantifying conceptual model uncertainty. Ground Water 48 (5), 701–715.

Toth, M., Royer, T., Peebles, R., Roth, M., 2013. Using analytics to quantify the value of seismic data for mapping Eagle Ford sweetspots. In: Unconventional Resources Technology Conference, Denver, CO, August 12–14.

Vapnik, V., 1995. The Nature of Statistical Learning Theory. Springer, New York.

Zhong, M., Schuetter, J., Mishra, S., Lafollete, R., 2015. Do data mining methods matter? A "Wolfcamp" shale case study. In: SPE Hydraulic Fracturing Technology Conference, Houston, TX, February 5–7.

Concluding Remarks

9.1 THE PATH WE HAVE TAKEN

9.1.1 Recapitulation of Topics

This book seeks to provide the background to understand and apply the fundamental concepts of classical statistics and emerging concepts from data analytics in the analysis of petroleum geoscience and related datasets. To that end, Chapter 1 started with definitions of statistics and data analytics, description of the data analysis cycle, overview of example applications in petroleum engineering and geoscience, and basic probability and statistics concepts. This was followed in Chapter 2 by various exploratory data analysis techniques for summarizing and visualizing univariate and bivariate data. Chapter 3 dealt with a number of common probability distributions and how to model them, along with concepts of confidence limits and comparing distributions. In Chapter 4, we looked at linear regression with two variables as a fundamental tool for modeling relationships, along with its extension for multiple variables and with nonparametric transformations. Multivariate analysis was covered in Chapter 5, which included topics such as dimensionality reduction, clustering, and discriminant analysis. Uncertainty quantification was the subject of Chapter 6, covering uncertainty characterization from empirical data or subjective judgment, uncertainty propagation using Monte Carlo simulation or analytic alternatives, and uncertainty importance (sensitivity) analysis using a number of techniques. Chapter 7 dealt with experimental design and response surface methods that included both classical and sampling-based designs. Finally, Chapter 8 focused on data-driven modeling methods, that is, those based on machine-learning approaches such tree-based models, boosting and bagging approaches, support vector machines, and artificial neural networks.

9.1.2 Style and Intended Use

As we have emphasized in the preface, this is a book on the application of statistics, written by practitioners, for practitioners. As such, we have tried to strike a judicious balance between statistical rigor and formalism and practical considerations regarding the fundamentals and applicability of various relevant concepts. We hope that petroleum engineers and geoscientists will come away with a solid understanding of the statistical concepts underpinning both basic and advanced topics, which will be reinforced by the worked problems and exercises. This minimalist approach with respect to mathematical treatment may not appeal to the purist, but our experience suggests it is sufficient for developing a proper appreciation of various techniques and algorithms discussed in the book. Also, the data-driven modeling chapter has been written with the geoscientist in mind, who is likely to be more interested in becoming a smart user of machine-learning algorithms rather than a programmer of such methods.

The material has been arranged in the form of a "how-to" manual or a ready reference guide for practitioners in the petroleum geosciences. It will also benefit engineers and scientists working in related subsurface domains such as hydrogeology, geologic carbon sequestration, and nuclear waste disposal. We hope it will be regularly used as a desktop companion by those routinely dealing with the acquisition, interpretation, analysis, and modeling of data from field tests, laboratory experiments, and/or computer simulations. We envisage the book to be utilized also as a textbook for an upper division or graduate-level course on statistical modeling and analysis with a geoscience flavor. As an alternative, a first course in basic statistics and geostatistics can be created by combining the first five chapters of this book with the material on geostatistical topics such as variography, kriging, and simulation.

9.1.3 Resources

The online resource for this book is https://www.elsevier.com/books-and-journals/book-companion/9780128032794. It contains the following:

- All of the datasets used in the book
- Excel files for several of the example problems
- Solutions to exercises at the end of each chapter
- GRACE, an open-source software for nonparametric regression (Chapter 4)
- E-FACIES, an open-source software for multivariate analysis (Chapter 5)
- E-REGRESS, an open-source software for experimental design and response surface analysis (Chapter 7)
- Miscellaneous scripts for executing some of the data-driven algorithms discussed in Chapter 8 using the open-source software *R*.
- Links to other relevant open-source software as discussed in various chapters

9.2 KEY TAKEAWAYS

9.2.1 Which Variables?

Today's datasets are larger than ever before, with this abundance of riches posing a challenge to the analyst in terms of which variables to focus on. For example, in the case of production data from an unconventional reservoir completed with horizontal wells containing multiple hydraulic fractures, there could be hundreds of independent (predictor) variables broadly grouped under (a) well geometry, (b) fracturing fluid, (c) fracture-treatment conditions, (d) chemistry of produced water, (e) geologic parameters, and (f) rock mechanics parameters. It is very likely that there will be a high degree of redundancy among these variables. There is also the possibility that the data can be partitioned into statistically homogeneous subgroups, with each subgroup having its own unique input-output relationships. This is where the process of unsupervised learning can be applied to reduce the dimensionality of the data by combining similar variables and creating clusters of data points that can be analyzed separately. Section 5.4 describes how multivariate analysis techniques such as principal component analysis, clustering, and discriminant analysis can be utilized for such a problem.

In terms of dependent (response) variables, it is sometimes useful to develop a new derived variable that contains more information than the primary variables themselves. For example, in the unconventional production case, the initial production rate normalized by the length of the horizontal well could be more effective in building predictive models. Similarly, in the analysis of seismic data, particularly when large volumes of data are acquired from permanently embedded sensors, the "onset time" could be more revealing than the seismic attributes themselves (Vasco and Datta-Gupta, 2016; Hetz and Datta-Gupta, 2017).

For example, in Fig. 9.1, the time-lapse survey maps of a two-way acoustic travel time shift because of steam injection in a heavy oil reservoir do not seem to contain any specific feature. However, the seismic onset time map, which is the calendar time when the seismic attribute crosses a prespecified threshold value at a given location, clearly seems to indicate a propagating front. Also, the onset time map reduces multiple time-lapse seismic survey data into a single map, leading to substantial data reduction.

9.2.2 Simple Model, or Complex?

Geoscientists are generally guided by Occam's razor, which states that a simpler explanation (model) is to be preferred over a complex one. This conventional wisdom can be at odds with a statistician's perspective, that is, the simplicity and interpretability of a model tend to be inversely related to its accuracy—leading to what can be described as Occam's dilemma. For example, in Section 8.4, the simple and easy-to-interpret ordinary least-squares model is found to be much less accurate than more complex and opaque models such as random forest or support vector machine. The challenge with embracing the latter alternative is how to tease out more information regarding the inner workings of the model. Techniques such as variable importance or conditional sensitivity can be useful tools in this regard (see Section 8.4.3). The modified conventional wisdom therefore suggests that the curse of dimensionality can be ignored without sacrificing interpretability for accuracy, by making sure that insights regarding variable interaction and input-output dependencies are extracted as part of the model-building process.

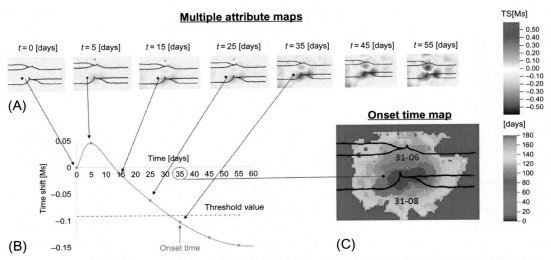

FIG. 9.1   **Conversion of multiple seismic attribute maps (time shift) to an onset time map. (A) A sample of seven time shift maps, (B) a plot of the seismic response at a specific cell (*black dot* in the time shift maps), and (C) the seismic onset time map derived from 175 time shift attribute maps. The contours (isochrones of onset times) display the front progression.**

9.2.3 One Model, or Many?

Machine-learning techniques such as random forest, support vector machine, and artificial neural networks are becoming increasingly popular for building input-output models as opposed to basic linear regression or its nonlinear/nonparametric variants. Often, the choice of which advanced model should be used is based on the analyst's preference. However, our experience suggests that no single method works best for all problems—making the *a priori* selection of a single modeling technique quite difficult. Sometimes, multiple competing models may arise when the goodness-of-fit measured in terms of training or test error is quite similar across the model set. Such an example is shown in Fig. 9.2, where the model fit expressed in terms of cross validation scaled RMSE is found to be very similar for four different models—ordinary kriging, quadratic fit with LASSO, multiple adaptive regression spline (MARS), or additivity and variance stabilization (AVAS). Here, the bars represent

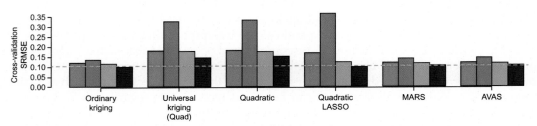

FIG. 9.2   **Model fits for multiple experimental design and response surface combinations.** *After Schuetter, J., Mishra, S., Zhong, M., LaFollette, R., 2015. Data analytics for production optimization in unconventional reservoirs. In: Proc. Unconventional Resources Technology Conference. DOI: 10.15530/URTEC-2015-2167005.*

different experimental design strategies, that is, orange, Box-Behnken (BB); purple, augmented pairs (AP); green, maximum entropy (ME); and black, maximin LHS (MM).

Compounding this problem even further, we have the possibility that each modeling approach provides different insights regarding the relative importance of the predictors. This was demonstrated earlier in Fig. 8.13. Our recommended solution is to accept the multiplicity of models and combine them using the process of model aggregation as discussed in Section 8.3.3. Aggregating over a large set of completing models provides more robust understanding and predictions compared with a single model, which may not be the most accurate for the problem at hand and may not capture the full range of variable interactions as part of model building.

9.2.4 Is Past Always Prolog?

A common question is whether data-driven modeling tools can replace physics-based models in dealing with subsurface processes. Here, it is important to recognize that statistical techniques have a limited ability to project the "unseen." The learning from sandstone formations cannot be directly applied to carbonates. Similarly, insights regarding early-time transient flow conditions during flow to wells do not carry over to late-time boundary-dominated flow conditions.

Consider Fig. 9.3, which shows the time-dependent rate-normalized pressure response in a tight-gas well (Palacio and Blasingame, 1993). Here, the flow behavior of the system can be clearly separated into the early-time transient period (shown by the quarter-slope lines) and the boundary-dominated period (shown by the unit-slope lines). What if we did not have any late-time data? In that case, with a physics-based model, multiple scenarios can still be generated for the late-time period even without any boundary-dominated data. However, with a data-driven model, only the past (transient) trend can be extrapolated into the future. It is therefore incumbent on the analyst to ensure that the conditions for which model predictions will be developed are consistent with those for the training dataset.

9.2.5 To Fit, or Overfit?

The flexibility of advanced statistical models such as those discussed in Chapter 8 can be both a blessing and a curse in the context of developing input-output models with good predictive accuracy. These models have much greater leeway for handling variable interactions, resulting in higher accuracy in general compared with linear regression or variants thereof. At the same time, there is the danger of overfitting to the training data by manipulation of the adjustable (tuning) model parameters (e.g., size of tree in random forest and number of hidden layers in artificial neural networks). This is where the power of cross validation (Fig. 8.11) can help the analyst. It can be used as an effective tool during the model-building process for balancing model accuracy and degree of complexity. Kuhn and Johnson (2013) have described such a workflow, which is summarized in Fig. 9.4. Also shown therein is an example of automated parameter tuning for a support vector machine model, and how the optimal value of the adjustable "cost" parameter is determined based on a cross-validated accuracy metric.

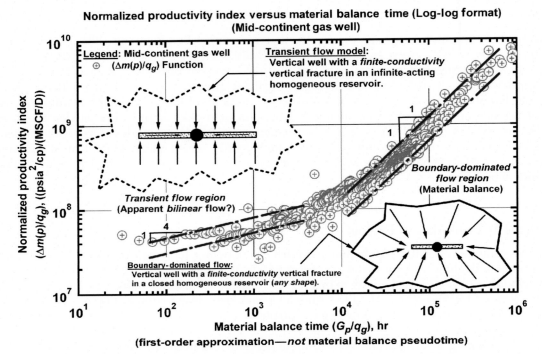

FIG. 9.3 Pressure and rate response for a tight-gas well showing both early- and late-time behavior. *After Palacio, J.C., Blasingame, T.A., 1993. Decline-curve analysis with type curves-analysis of gas well production data. Paper SPE 25909 Presented at the SPE Rocky Mountain Regional/Low-Permeability Reservoirs Symposium, 12–14 April, Denver, SPE 25909-MS. https://doi.org/10.2118/25909-MS.*

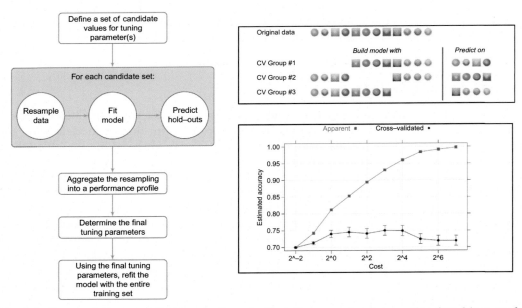

FIG. 9.4 Process for automated determination of adjustable parameters in complex data-driven models. *After Kuhn, M., Johnson, K., 2013. Applied Predictive Modeling. Springer.*

9.3 FINAL THOUGHTS

In summary, we note that there is a growing trend toward the use of statistical modeling and data analytics for oil and gas (and related subsurface domain) applications. The goal is to "mine the data" and develop data-driven insights to understand and optimize the performance of geosystems such as petroleum reservoirs and groundwater aquifers for both fluid production and injection problems. The maturity of the field appears to be much like that of geostatistics in the early 1990s, when it had not been fully adopted for mainstream applications. To that end, we believe that petroleum engineers and geoscientists need to develop a better understanding of the full repertoire of available techniques and their potential. This will help them better interact with data scientists to propose and apply appropriate statistical techniques for developing robust data-driven insights for decision-making.

We close with a quote from the poet T.S. Eliot, who asked "Where is the wisdom we have lost in knowledge? Where is the knowledge we have lost in information?" Let these thoughts guide our journey as we strive to transform data into information by understanding relations, transform information into knowledge by understanding patterns, and transform knowledge into wisdom by understanding principles (Bellinger, 2004).

References

Bellinger, G., 2004. Data, information, knowledge and wisdom, http://www.systems-thinking.org/dikw/dikw/htm.

Hetz, G., Datta-Gupta, A., 2017. Integration of continuous time lapse seismic data into reservoir models using onset times. In: First EAGE Workshop on Practical Reservoir Monitoring, March 6–9, 2017, Amsterdam.

Kuhn, M., Johnson, K., 2013. Applied Predictive Modeling. Springer, New York.

Palacio, J.C., Blasingame, T.A., 1993. Decline-curve analysis with type curves-analysis of gas well production data. Paper SPE 25909 Presented at the SPE Rocky Mountain Regional/Low-Permeability Reservoirs Symposium, 12–14 April, Denver, SPE 25909-MS. https://doi.org/10.2118/25909-MS.

Schuetter, J., Mishra, S., Zhong, M., LaFollette, R., 2015. Data analytics for production optimization in unconventional reservoirs. In: Proc. Unconventional Resources Technology Conference. https://doi.org/10.15530/URTEC-2015-2167005.

Vasco, D.W., Datta-Gupta, A., 2016. Subsurface Fluid Flow and Imaging: With Applications for Hydrology, Reservoir Engineering and Geophysics. Cambridge University Press, London.

Index

Note: Page numbers followed by *f* indicate figures, *t* indicate tables, and *b* indicate boxes.